Flower Crop Cultivation and Management

Flower Crop Cultivation and Management

Edited by:

Mohan Balasaheb Patil

Delve
PUBLISHING

www.delvepublishing.com

Flower Crop Cultivation and Management

Mohan Balasaheb Patil

Delve Publishing
224 Shoreacres Road
Burlington, ON L7L 2H2
Canada
www.delvepublishing.com

Email: orders@arclereducation.com

ISBN: 978-1-77407-654-5 (Hardcover)

ABOUT THE EDITOR

Dr Mohan Balasaheb Patil is presently working as Professor and In charge of fruit research station Aurangabad. He did his Bsc Agri. from Marathwada Agril. University Parbhani in 1986. Msc (Ag) Horti. In 1988 while, Ph. D in 2004 from the same university. He has started his carrier in 1994 as Asistant Prof. promoted as Associat Prof. in 2007 & Professor in2017. He has established the Sweet Orange Research Station at Badnapur. Published 25 research papers in national and international journals. Published more than 100 popular articles. He reeived Agrocare Award for the best extension activities. He has authored and published two books.

TABLE OF CONTENTS

List of Figures .... *xi*

List of Tables .... *xix*

Preface .... *xxi*

Chapter 1 Introduction To Flower Crops Cultivation **1**

1.1 Introduction 2

1.2 Benefits of Flower Crop Cultivation 3

1.3 Types of Flower Crops Growing Methods 3

1.4 Categories of Flower Crops 8

1.5 Economic Importance of Flower Production 17

1.6 Global Floriculture Outlook 19

1.7 Career Paths In Flower Crops Production 21

Chapter 2 Major Types of Flower Crops **27**

2.1 Introduction 28

2.2 Bedding Flowers 29

2.3 Flowers 36

2.4 Potted Flowers 41

2.5 Indoor Foliage 46

2.6 Cut Flowers 52

Chapter 3 Tasks Involved In Flower Crops Cultivation **61**

3.1 Introduction 62

3.2 Deciding What Flowers To Grow And Organizing Their Planting 62

3.3 Preparing Soil For Planting Flowers 64

3.4 Maintaining Flowers Which May Involve Spraying, Pruning, Fertilizing and Watering 66

3.5 Grading And Packing Flowers For Sale 68

3.6 Marketing And Selling Flowers 70

3.7 Transporting Flowers To Buyers or Exporters 73
3.8 Maintaining And Repairing Machinery Such as Pumps and Other Equipment 75
3.9 Hiring and Training Workers, and Ensuring That Their Work Is of a High Standard In Flower Farms 77
3.10 Ensuring That Health And Safety, And Other Regulations, are Complied With In Flower Farms 79
3.11 Ensuring Market Certification Systems (Such As Quarantine Requirements) Are Up To Date 81
3.12 Keeping Production And Financial Records And Write Reports 83
3.13 Researching Growing And Harvesting Techniques, and New Varieties 85

Chapter 4 Propagation Methods of Flower Crops 89
4.1 Introduction 90
4.2 Raising Flower Crop From Seeds 91
4.3 Raising Flower Crop Through Vegetative Propagation 103
4.4 Raising Flower Crop Through Tissue Culture 109

Chapter 5 Cultivating Flower Crops In Greenhouses 111
5.1 Introduction 112
5.2 Advantages Of Growing Flower Crops in Greenhouses 113
5.3 History Of Growing Flower Crops in Greenhouses 114
5.4 Growing Systems For Flower Crops in A Greenhouse 116
5.5 Characteristics Of Flower Growing Medium in a Greenhouse: 124
5.6 Important Flower Crops Commercially Grown in Greenhouses........ 133

Chapter 6 Challenges And Constraints In Flower Crops Cultivation 141
6.1 Introduction 142
6.2 Deficiency Information on New Varieties of Flower Crops 142
6.3 Lack of Precise Information on The Production of Different Flower 143
6.4 Poor Transportation 144
6.5 Lack of Organized Marketing Co-Operatives 145
6.6 Lack of Research And Development In Floriculture 146
6.7 Lack of Market Intelligence 147
6.8 Lack of Detailed Economic Feasibility Studies 147
6.9 Lack of Streamlined Quality Control Mechanism 148

6.10 Poor Co-Ordination Between Government And Private Agencies Involved In Import And Export of Floriculture Products 149
6.11 Lack of Information on Advanced Cultivation Practices 150
6.12 Lack of Appropriate Planting Materials 150
6.13 Lack of Knowledge About Importing And Exporting Countries 151
6.14 Lack of Quality Consciousness of The Produce 152
6.15 Lack of Sale Promotion Activities 153

Chapter 7 Pests And Disease Affecting Flower Crops 155
7.1 Introduction 156
7.2 Insects Causing Damage by Chewing 156
7.3 Insects Causing Damage by Piercing-Sucking 162
7.4 Mining Insects 166
7.5 Boring Insects 167
7.6 Major Diseases of Flower Crops 171
7.7 Management of Pests And Diseases of Flower Crops 183

Chapter 8 Management Practices For Growing Flower Crops 185
8.1 Introduction 186
8.2 Environmental Factors For Growing Flower Crops 186
8.3 Soil And Growing Medium For Flower Crops 192
8.4 Optimum Spacing and Planting Density Of Flower Crops 195
8.5 Budding And Pruning Flower Crops 196
8.6 Optimum Nutrients For Flower Crops 199
8.7 Pinching Flower Crops 201
8.8 Disbudding Flower Crops 203
8.9 De-Shooting Flower Crops 205
8.10 Defoliation Flower Crops 206
8.11 Removal of Faded Flowers 208
8.12 Bending of Shoots Of The Flower Crops 210
8.13 Irrigating Flower Crops 211
8.14 Disease Control In Flower Crops 213
8.15 Insect And Pest Control In Flower Crops 216
8.16 Weeding 218

Chapter 9 **Harvest and Post-Harvest Handling Of Flower Crops 221**

9.1 Introduction 222

9.2 Pre-Harvest Factors Affecting Flower Longevity. 223

9.3 Harvest Factors Affecting Flower Longevity 225

9.4 Post-Harvest Factors Affecting Flower Vase Life 228

9.5 Harvesting Flowers 239

9.6 Flower Storage Methods 240

9.7 Storage Conditions For Flower Crops 244

9.8 Grading And Sorting Flowers 247

9.9 Bunching 248

9.10 Chemical Solutions 249

9.11 Packing 251

9.12 Cooling 253

9.13 Irradiation 258

Chapter 10 **Sustainability In Flower Crop Cultivation 261**

10.1 Introduction 262

10.2 Factors Affecting Sustainability In Flower Crop Cultivation 262

10.3 Benefits Of Sustainable Flower Crop Cultivation 267

10.4 Initiatives To Maintaining Sustainability In Flower Cultivation. 273

References 281

Index 285

LIST OF FIGURES

Figure 1.1: Flower growing is a fulfilling activity that can be done for both recreational and commercial purposes

Figure 1.2: Flowers cultivated in a protected environment are safe from harsh weather conditions and other unfavorable external elements

Figure 1.3: Flowers grown under field conditions

Figure 1.4: Illustrating bedding flowers

Figure 1.5: Examples of cut flowers

Figure 1.6: Illustrating examples of potted flowers

Figure 1.7: Flowers are widely used in the commercial production of perfumes and fragrances

Figure 1.8: Preparation of floral medicine

Figure 2.1: These are pink petunia flowers.

Figure 2.2: This shows the high diversity in the colors of the Busy Lizzie.

Figure 2.3: Geraniums grow in clusters on one stem.

Figure 2.4: Lobelias mostly come in lilac as seen above.

Figure 2.5: The black eye with a cone-shaped head of the rudbeckia is their main determining factor.

Figure 2.6: They have long slender stems with a cluster of flowers per stem.

Figure 2.7: They have a characteristic appertaining as they hang on the length of the stem downwards. They have a unique heart-shaped flower.

Figure 2.8: The center of the flower is shaped like a cone, giving them their name coneflower.

Figure 2.9: Their leaves being purple can be mistaken for petals. They give a vibrant color that attracts the eye.

Figure 2.10: Their flowers may release a very fragrant smell.

Figure 2.11: The bright colors of this flower allow it to look radiant indoors.

Figure 2.12: They have a distinct trumpet-shaped flower.

Figure 2.13: In many places, these are a sign of luck. They are associated with St. Patrick'sDay.

Figure 2.14: This clearly shows their upswept flowers and pattered leaves.

Figure 2.15: They have unique long slender leaves. The flowers may be mistaken for leaves due to their orientation.

Figure 2.16: Ferns do not produce flowers. They reproduce using spores.

Figure 2.17: Their leaves have yellowish streaks that are pleasant and decorative.

Figure 2.18: The leaves are variegated with white and are heart-shaped. They could also have yellow or pale green variegations.

Figure 2.19: These can grow on trees and into canopies in the wild.

Figure 2.20: Wandering Jew plant.

Figure 2.21: These are bred for their shape, size, color and fragrant smell. They are the most common cut flower type.

Figure 2.22: The flowers are tiny but they compensate for that in their volume. They can produce hundreds of flowers per plant.

Figure 2.23: These are large-flowered carnations. They have one large flower per stem. They also come in many different colors and some even bicolor.

Figure 2.24: Bicolor lily with a white border and dark spots. Lilies come in many shapes, colors and sizes.

Figure 2.25: They are very similar to lilies having trumpet-shaped flowers. They, however, grow from tuberous roots.

Figure 2.26: Mums have many petals on one flower giving them their distinct cotton ball look.

Figure 3.1: Flower growers should choose the best breed to cultivate based on factors like soil pH, growth rate and ease of maintenance

Figure 3.2: Soil preparation is an important step in flower crops cultivation

Figure 3.3: Applying compost or granular organic fertilizers will feed your soil and plants with nutrients, which will lead to healthy plants

Figure 3.4: Flower grading and packaging is a delicate process that requires knowledge and accuracy

Figure 3.5: Marketing and selling of flowers is an important task to flower cultivation

Figure 3.6: Transporting flower products from one point to another requires adequate care, and without it things may become pretty messy

Figure 3.7: Equipment like ventilation, lighting and watering systems found in a flower garden should be well-maintained to ensure quality yield

Figure 4.1: Propagation is a biological process used by growers to reproduce and multiply their flowers

Figure 4.2: Stratification involves putting seeds in frozen conditions so that they can germinate quickly once introduced to warm temperatures again

Figure 4.3: Soaking is an effective way of softening hard flower seeds in preparation for planting

Figure 5.1: Victoria Amazonica in large greenhouse at the saint Petersburg botanical garden.

Figure 5.2: Greenhouse production

Figure 5.3: The royal greenhouses of laeken, a masterpiece of 19th century.

Figure 5.4: ground bed preparation.

Figure 5.5: Raised bed gardening

Figure 5.6: Ebb and flow type benches

Figure 5.7: Container flower gardening using pot

Figure 5.8: The greenhouse provides a safe and controlled environment for flower growing unlike in field planting

Figure 5.9: Soil can be categorized based on its water holding properties

Figure 5.10: The rose flower is popularly grown in greenhouses for aesthetic purposes

Figure 5.11: Chrysanthemum flower in greenhouse

Figure 5.12: Tulip flower in greenhouse

Figure 5.13: Gypsophila in a greenhouse

Figure 6.1: A rural Columbian farmer transports his flower produce to market through poor roads that can flood after a downpour

Figure 6.2: Constant research is necessary for developing and improving flower varieties

Figure 6.3: A flower expo in Poland showing quality flower strains shipped from different countries

Figure 7.1: Caterpillars are moth and butterfly larvae that feed of foliage

Figure 7.2: Transparent maggot pests feeding on a flower crop

Figure 7.3: Beetle chewing on fresh flower

Figure 7.4: Sawfly larvae hanging onto a flower for food

Figure 7.5: Grasshoppers have sharp incisors on their mandibles used to tear down crops

Figure 7.6: Slugs and snails may appear similar but snails have a shell for differentiation

Figure 7.7: A colorful leafhopper insect with serrated appendages

Figure 7.8: Thrips can be identified by their yellowish-brown body and tough exoskeleton

Figure 7.9: Aphids have a rounded-body and long slim legs

Figure 7.10: Mites have a hairy body and a set of eight legs also covered with hair

Figure 7.11: A leafminer sinking its mandibles into foliage, they usually create spiral blotches on leaves

Figure 7.12: Stalk borers usually live in the stalks of infested plants

Figure 7.13: With its distinctive mosaic-patterned wing, the burdock borer moth is quite easy to detect

Figure 7.14: A pink colored iris borer destroying flowers

Figure 7.15: A European corn borer with its triangular orange-brown wings

Figure 7.16: Crown gall rot makes plants very unsightly to look at

Figure 7.17: Blackspot fungi causes dark blemishes on flower leaf surfaces

Figure 7.18: Powdery mildew causes plants to appear frost-like

Figure 7.19: Downy mildew spreading across a garden

Figure 7.20: Leaf rust is caused by fungi present in the environment

Figure 7.21: Anthracnose can make leaves lose their capacity to conduct photosynthesis

Figure 7.22: Grey mold can form both on leaves and the soil surface

Figure 7.23: This disease makes crops to appear weak and dreary

Figure 7.24: Sooty molds often turn leaf surfaces from green to black in color

Figure 7.25: Canker lesions devastating a piece of leaf

Figure 7.26: Rose mosaic disease affecting a germinating rose flower leaf

Figure 7.27: A wilting rose plant with fading flowers

Figure 7.28: Rose rosette eats up flowers making them to fade away

Figure 7.29: Frost can destroy flowers by freezing their water content

Figure 7.30: Excess salinity can affect the growth of roses

Figure 7.31: Glyphosphate is one of the dangerous chemicals found in herbicides. It can deform flowers.

Figure 8.1: This shows the effect on low temperatures on the leaves of this potted plant. The leaves have been discolored and turned brown.

Figure 8.2: This shows how extreme light causes fading of flowers.

Figure 8.3: These are some of the effects of excessive or too little humidity.

Figure 8.4: CO2 is very important in the growth of flowers for without it there is no photosynthesis thus flowers cannot be produced by the plant.

Figure 8.5: Different flower species prefer different soils to grow appropriately. The different soils have different properties.

Figure 8.6: This is rockwool growing medium. It is made from melted rocks and has very good water retention capabilities.

Figure 8.7: Perennials should be spaced according to how wide they grow. The distance should be measured from the center of one plant to the center of the other.

Figure 8.8: Budding is the process of taking a bud plant and growing it on another plant. The bud grows to be similar to the plant from which it came from.

Figure 8.9: This shows the gardener cutting of the spent flower so as to improve the aesthetic value of the blooms. The spent flower has withered petals and that makes the cluster inconsistent.

Figure 8.10: Eggshells can be used to provide nutrients to the soil as a form of compost. They decompose and release nutrients.

Figure 8.11: Earthworms are very beneficial in improving soil nutrition. They feed on materials in the soil that are either dead or decaying and thereafter excrete dung that is rich in nutrients.

Figure 8.12: Pinching is very beneficial to flower plants. This is an example of a spent flower being pinched out at the stem. This provides bigger blooms.

Figure 8.13: Disbudding helps in providing a larger bloom and making the bouquet look fuller. The central bud should be left intact as seen above .

Figure 8.14: Removing shoots can help increase the number of flowers that grow per shoot, giving a fuller bouquet.

Figure 8.15: Leaves that are discolored and browned should be defoliated so as to make the foliage more attractive.

Figure 8.16: This is the Schwazzing technique which involves removing fam leaves to allow light to reach the buds below.

Figure 8.17: Faded flowers should be removed so as to allow the plant to use its energy on improving the health of the other flowers and the plant itself.

Figure 8.18: Bending helps increase the leaf area, allowing the plant to build a strong root system for acquiring nutrients for the plant.

Figure 8.19: Sprinklers are not advisable in irrigating flowers because they wet the foliage, making the plant susceptible to diseases and pests.

Figure 8.20: This fungus can spread on the whole leaf appearing as a white powdery layer.

Figure 8.21: Wetting the flower plants foliage can lead to the growth of thus fungus. It can be very problematic during the rainy season.

Figure 8.22: Aphids attack a rose flowers leaving it looking very unattractive and it can kill the flower crop.

Figure 8.23: Japanese beetles feed on flowers leaving them looking like unattractive.

Figure 8.24: Weeds compete with flower crops for nutrients water and even light thus having lower flower yields and making the plants susceptible to diseases and pests.

Figure 9.1: Harvesting is best done in small quantities at a time and the gardener should ensure that the flowers are immediately taken to a cool area to maintain their quality.

Figure 9.2: The relative humidity is affected by changes in temperature. A rise in temperature can cause condensation which wets the surfaces of the flowers and leaves.

Figure 9.3: This shows the difference in harvesting lavender to late and at the right time.

Figure 9.4: Flowers should be harvested in small quantities at a time to avoid mechanical damage.

Figure 9.5: These roses should be harvested when the first two petals begin to unfold and the calyx is reflexed below a horizontal position

Figure 9.6: Some flowers are harvested at the bud stage and others when they are fully open.

Figure 9.7: condensation on flowers can lead to growth of mold and fungi reducing flower quality.

Figure 9.8: this shows how water is lost by plants through transpiration.

Figure 9.9: This is an example of a floral preservative. It contains sugar or dextrose, an antibacterial agent and an acidifier

Figure 9.10: This is the effect of Ethylene on flowers. It causes fading and loss of color of flower petals.

Figure 9.11: This is positive tropism shown by the plants growing towards the light. Phototropism.

Figure 9.12: Harvesting is best done in small quantities at a time to prevent mechanical damage by bruising.

Figure 9.13: Botrytis blight is the most common disease of cut flowers.

Figure 9.14: One should have enough clean buckets before cutting flowers. The buckets and pruners should be well cleaned with a tablespoon of bleach per gallon of water.

Figure 9.15: Refrigerators must have precise temperatures for the flowers they are storing to ensure maximum flower quality. The ideal temperature for a flower refrigerator is between 1.11 to 2.22 degrees Celsius.

Figure 9.16: This storage method works by decreasing the level of oxygen and increasing the concentration of CO2 by controlling the gas composition precisely.

Figure 9.17: This method has two disadvantages. The cost of installation is too high and it causes water loss from the tissues.

Figure 9.18: Flowers need to be kept in cool areas instead. If the storage temperature is high, the rate of respiration will increase and so will water loss.

Figure 9.18: The relative humidity required by flowers is around 75-95%.

Figure 9.20: There are grading stations with a grader at each. The grader takes the ungraded cut flower material and grades and simultaneously bunches the cut flower material in fixed, decision limiting racks.

Figure 9.21: Strings or elastic bands are used to hold together the bunches.

Figure 9.22: This shows the results of rehydration of flowers.

Figure 9.23: Budding solution improves the balance of water and also improves the longevity of the flowers.

Figure 9.24: Cardboard can become very inconvenient in areas with high humidity due to loss of strength that may cause the box to collapse. One needs to ensure the box material is of good quality.

Figure 9.25: Overpacking boxes should be avoided so as to maintain the pre-cooling effect.

Figure 9.26: According to the diagram, cooling would be dome for more than two hours to reach the temperature of the cooling air.

Figure 9.27: These are examples of tropical flowers which are of high market value.

Figure 9.28: These flowers are packed vertically in boxes that have water with a biocide in the bottom

Figure 9.29: This method works on the basis of water cools at a lower temperature when boiled at a low atmospheric pressure.

Figure 9.30: This process is done by passing packed flowers through a radiation chamber on a conveyor belt.

Figure 10.1: Soil nutrients are important for flower crops cultivation

Figure 10.2: Investing in irrigation methods that are low pressure and low volume such as drip irrigation and the use of micro-sprayers can help minimize water use.

Figure 10.3: Destroying soil cover can lead to a rise in pests and disappearance of beneficial animals, insects and micro-organisms thus lowering quality of flower crops.

Figure 10.4: Diversifying the flowers planted and grown on the farm is a risk management strategy.

Figure 10.5: Soil is a natural resource that we cannot renew yet we are literally throwing it away.

Figure 10.6: This is the principle of Equity and Equality in social and economic justice.

Figure 10.7: Training the farmers about the business principles inclusive also of planning and management can go a long way in improving flower quality and sustainability.

Figure 10.8: Agrochemicals can affect the health of farmers and any people living in the affected areas. Excessive and improper use of fertilizers can contaminate water sources.

Figure 10.9: This is a drought resistant perennial that can be planted ahead of droughts to minimize water use.

Figure 10.10: This shows the chain of transparency.

Figure 10.11: Wind can be used to generate electricity and also to pump water. Turbines can be installed on flower farms to provide electricity.

Figure 10.12: Aphids sack the sap out of plant leaves and also feed on the leaves leaving them looking very unattractive and it can kill the plant.

LIST OF TABLES

Table 9.1: These are some optimum temperatures for storage if different flowers.

PREFACE

Flower crop cultivation is regarded as one of the modern farming techniques that havethe potential to spur production and is increasingly receiving attention in many countries. In developing countries, flower crop cultivation is a potential source for economic growth and industrial development. Normally it is a growing sector, and therefore, it is one of the major sources of employment for domestic industries, and for earning foreign exchange. Primarily, this sector forms the basic potential source of savings for agricultural and nonagricultural investment. For these vital roles to be fulfilled, high rates of output growth must be sustained.

There are very few books that provide the necessary information on flower crop production to farmers. This book covers materials that are extremely useful for the day-to-day operation of a farm or garden. It contains materials that take into account all the social and physical/environmental factors revolving around flower crop cultivation.

Despite the great milestone in the growth of floriculture sector in the last few decades, strongly driven by improved fertilizer application, modern irrigation schemes, the use of agricultural machinery, pesticides application, and proper land management, it would be indefensible to assume that these relationships will remain linear in the future. The book is primarily designed for use by all undergraduate and postgraduates studying floriculture. Extension workers, agricultural researchers, flower specialists, horticulturists, and other stakeholders will also find the book to be useful. It can also be used to train and equip small-scale flower farmers and gardeners in the agricultural sector. It is an adventure worth taking.

For the convenience of readers, the whole book is divided into ten chapters. Chapter 1 deals with an introduction to flower crops cultivation; chapter 2 discuss the major types of flower crops; chapter 3 deals with tasks involved in flower crops cultivation; chapter 4 focuses on the methods of flower crop propagation; chapter 5 deals with cultivating flower crops in greenhouses; chapter 6 unravels the challenges and constraints in flower crop cultivation; chapter 7 interestingly casts light onthe pests and diseases affecting flower crops; chapter 8 is on the management practices for growing flower crops; chapter 9 is on the harvest and post-harvest handling of flower crops; and chapter 10 discuss the sustainability offlower crop cultivation. All the 10 chapters are presented briefly with an introduction section. Detailed information is offered in the other sections.

I sincerely believe and hope that the information presented in this book will be useful to you.

1

INTRODUCTION TO FLOWER CROPS CULTIVATION

CONTENTS

1.1 Introduction 2

1.2 Benefits of Flower Crop Cultivation 3

1.3 Types of Flower Crops Growing Methods 3

1.4 Categories of Flower Crops 8

1.5 Economic Importance of Flower Production 17

1.6 Global Floriculture Outlook 19

1.7 Career Paths In Flower Crops Production 21

1.1 INTRODUCTION

Flower growing, also known as flower farming or floriculture, is the planting of flowers or ornamental flora for gardens, décor or floristry. The creation of new plant varieties through controlled breeding is yet another key role of floriculturists.

Some common floriculture vegetation includes the following:

- Bedding plants
- Potted flower
- Houseplants
- Cut flowers

Different from nursery plants, floriculture crops are largely herbaceous. For bedding and garden vegetation, they mostly comprise of annuals and perennials as young flowering plants, including vegetable plants. These are cultivated in cell packs such as flats or trays, pots or sometimes even in hanging baskets within a controlled environment. These flowers are sold mostly for gardens and landscaping purposes, with Impatiens, Petunia and Pelargonium varieties regarded as the most popular bedding plants. In the U.S., the diverse cultivars of Chrysanthemum form the main perennial garden plant breed. Being delicate products, flower growing requires special skills in spacing, nurturing, and pruning for optimal development. Equally, post-harvest treatment for cut flowers is important and requires chemical treatments, proper storage, preservation and packaging. So that the flowers can be delivered to the final user in top condition (USDA, 2019).

Figure 1.1: Flower growing is a fulfilling activity that can be done for both recreational and commercial purposes.

Source: https://www.gardeners.com/how-to/growing-annual-flowers-from-seed/5663.html

1.2 BENEFITS OF FLOWER CROP CULTIVATION

Flower growing has numerous health benefits, especially when done in your garden.

- It can help to increase your feelings of well-being and uplift the spirit
- For most people, being around nature elicits positive senses that generally make them happier.
- Flowers and gardening plants provide some form of eco-therapy that treats the mind.
- Regular nurturing activities can help keep you physically active and improve your fitness levels.

1.3 TYPES OF FLOWER CROPS GROWING METHODS

1.3.1 Flowers Grown Under Protected Environment

In this model, a covering structure which is usually transparent is used to protect the plants from harsh environmental factors. The flora is grown in a form of greenhouse construction that's been curated to develop a favorable micro-climate. Since the growth conditions are manageable, cultivation could be possible throughout the year or certain parts of the season as required. Even in cold climate zones, where outdoor planting is otherwise impossible, under-protected structures, healthy flowers can still be brought up.

The primary environmental factor traditionally managed is temperature, usually, the protective layer provides warmth to overcome excess cold conditions. Still, environmental control may include cooling to alleviate excessive temperatures, control lighting conditions, carbon dioxide amounts, relative moisture, water, soil nutrients and vermin control (USDA, 2019).

Figure 1.2: Flowers cultivated in a protected environment are safe from harsh weather conditions and other unfavorable external elements.

Source: https://www.shutterstock.com/video/clip-1014353432-large-glass-greenhouse-flowers-growing-greenhouses-interior

There are different forms of protected flowering structures:

Low tech greenhouse: This is a simple growing structure built with locally accessible materials like bamboo or timber. Usually, an ultraviolet (UV) layer is also applied for cladding materials. Different from typical or hi-tech greenhouses, there are no particular control devices for modifying environmental factors within the greenhouse. Basic techniques are, however, utilized for increasing or reducing the temperature levels and moisture. Even light amounts can be minimized by integrating shading materials such as nets, plus the heat can be brought down during the summer season by simply opening the side-walls. Alternatively, in cold weather the internal temperature may be increased by covering the sidewalls using a plastic film. This form of a greenhouse is mostly ideal for cold climatic zones.

Medium-tech greenhouse: Some flower growers prefer having manually or semi-automatically controlled structures due to the low investment required. This kind of greenhouse is built using galvanized iron tubes. The canopy layer is connected with the main structure through screws, while the entire structure is solidly fixed onto the ground in such a way that it can withstand disturbances caused by heavy wind.

Exhaust fans having a thermostat function may be provided to regulate the temperature. Likewise, evaporative cooling pads or misting arrangements can be made to sustain favorable humidity levels within the greenhouse.

Since the systems are semi-automatic, thus requiring lots of attention and upkeep, they are mostly suitable for growing flowers in dry and composite weather areas. In some cases, flowers may also be cultivated in hi-tech protective structures, where the entire structure and environment parameters are controlled automatically.

Most flowers reared under protected conditions are watered through the micro-irrigation system, which experts believe is the best technique for watering flora within a greenhouse environment. Either micro-sprinklers or drip irrigation equipment may be applied for irrigation. Generally, the idea is that irriation systems must ensure the water doesn't fall directly on the flowers or leaves since this leads to infection and scorching issues. As for micro sprinkler units, water under great pressure is pumped through nozzles fixed on a supporting podium, at an elevation of around 1 foot high. Thus promotes watering at the bottom section of the plants. Some useful equipment needed for drip irrigation under-protected systems includes; a pump for generating pressure, water filtration unit and PVC piping with drippers (USDA, 2019).

1.3.2 Flowers Grown Under Semi-Protected

Semi-protected flower farming greenhouses provide energy efficiency while ensuring the grower gets a better yield for their efforts. The potential of this system in regards to microclimate control, plant protection, behavior, flower produce and energy savings is phenomenal. In fact, studies show that semi-closed systems have a more stable internal climate that's ideal for plants to grow in. They provide sustainable air-temperature and moisture compared to the fully closed or open growing structures, which are largely influenced by external atmospheric temperature and variable wind speeds.

With these units, it's also possible to achieve optimumCO_2 levels inside the greenhouse without any extra supply. Furthermore, use of air-cooling through pad or fan systems may be practiced to make the flowers more vigorous in summer seasons, through lowering of temperature and vapor pressure shortage easily. Controlled air-cooling mechanisms may also be useful in managing pre-night weather changes, which are otherwise unfavorable to flowers. Plants may not cope with varying day and night time temperatures, but with a semi-protected covering, they shall be able to manage. Furthermore, the capacity to maintain favorable heating set-points from lowered-temperature water in a semi-protected greenhouse is quite interesting, since it helps to reduce energy consumption (USDA, 2019).

Hydroponic flower growing: The adoption of hydroponic semi-protected flower structures has rapidly increased over the last few years. Hydroponic flower growing technique that uses water and controlled plant nutrients for cultivation ingredients, without soil. It's a model for high volume flower production, generating up to 9 times more flower production than the traditional field growing techniques while using just 20 percent of the water.

- Apart from production efficiency, hydroponic semi-protected structures provide greater levels of climate management, fertigation and plant protection, while requiring minimal resource inputs compared to soil-based flower farming. Furthermore, semi-covered hydroponic cultivation has a considerably clean ecological footprint, since CO_2 generated from burning organic gas in the green-house is retained within it. The air is usually recycled but not released into the environment like most typical glasshouse operations.
- All these factors, including CO_2 and heat retention, fewer chemical sprays, changing waste plant matter to produce heat, methane and manure as by-products give flower growers the opportunity to increase yield exponentially. While it may require some initial investment, this growing method still provides a great opportunity for cultivators to modernize their production techniques by reducing inputs and waste, therefore leading to growth and greater profitability for commercial growers. Additionally, variations between growth conditions in different localities suggest that some understanding of regional weather patterns, alongside with greenhouse form and structure are crucial aspects in determining whether a fully closed or semi-protected greenhouse is best. Also, energy utilization for heating or cooling will depend on local weather patterns, but in general this flower cultivation method is more suitable for those living in hot environmental conditions.

Various elements should be considered before a semi-covered greenhouse is used for flower growing. A few of these include; light conditions, temperatures, terrain and humidity levels which are normally experienced in your area. Still, nowadays, thanks to innovation items like air-conditioners and artificial lighting can be introduced to the flower greenhouse to make it more conducive (Thippaiah, 2005).

1.3.3 Flowers Grown Under Field Conditions

Open field cultivation is when the flowers are grown outside in a garden or farm setting, exposed to all the natural weather elements. In this arrangement, the plants are allowed to mature under normal climatic conditions where they're well-suited, within the region's prevailing ecological conditions.

Figure 1.3: Flowers grown under field conditions.

Source: https://www.doi.gov/blog/wild-wildflowers

1.3.4 Benefits of Field-based Flower Growing

- During suitable seasons, this flower planting method allows for simple and cost-effective streamlining for soil-preparation, sowing, fertilization and pesticide treatment. Additionally, for commercial horticulturalists, open-field production can be highly profitable since it allows for larger fields to be cultivated. As compared to protected or semi-protected zones, where the square footage available for planting is limited.
- An open field is less costly to manage, in contrast to protected flower production that requires lots of materials to put up. For the latter, growers need to buy various items such as heaters, lights, peat soil, fertilizers, and covering plastic or glasshouse. However, flowers cultivated in open-air environments don't require any special resources or accessories to grow. Just regular farming is enough and this significantly lowers the costs of production.

The biggest risk of protected flower growing is that it needs a huge investment which typically won't bring much return during the first few months, particularly for commercial flower farmers. They may put in a lot of resources into the project, but end up not getting enough returns for all their investments in installing a functional greenhouse for their flowers. Therefore, if some calamity or disease ends up destroying the plants after just a few months of growing, the cultivator may incur a huge loss. In comparison, even if an open field is affected by similar circumstances, the destruction won't be as costly as in the greenhouse example (Thippaiah, 2005).

Additionally, light is among the biggest challenges when growing flowers indoors. It's true that all plants require light in order to conduct photosynthesis, the process used for making flowers and food. However, most vertical farms or green-houses don't get sufficient sunlight, as such they supplement it with "artificial light" which is costly to maintain and usually not energy efficient. A generator may be required to provide electricity for lighting, but this is usually quite expensive to manage. Not to mention the immense carbon footprint that petrol generators used for lighting normally leave behind.

Some common artificial lights applied in indoor flower farms are gas-discharge lamps and fluorescent tubes, though lately energy-efficient LEDs have been preferred also by some farmers, particularly in growth cabinets. They may be low-energy alternatives to sunlight, but growing flowers under constant lighting or water pumping systems within growth chambers or greenhouses may require phenomenal amounts of energy. All the lights, glowing collectively within a closed system, will generate substantial amounts of heat and to maintain the optimal temperature for indoor plant farming more energy shall be required. Upon adding all these factors together, you are left with a huge carbon footprint and financial cost for each flower grown indoor. On the contrary, flowers cultivated under open farm conditions only need natural sunshine, which is a free, clean and renewable energy source (Thippaiah, 2005).

1.4 CATEGORIES OF FLOWER CROPS

1.4.1 Bedding Flower

Bedding, in flower growing, refers to the provisional planting of quick-growing flora into flower beds, for purposes of creating colorful, short-term, seasonal displays, which are used during spring, summer and winter seasons. Most bedding plants are usually annuals, biennials and tender perennials; though succulents are also increasingly gaining popularity. Some bedding floras are also known as "patio plants" since they are broadly used in pots, including other containers placed on patios, decking, terraces, or other zones around houses. Bigger tender "conservatory plants" can also be transferred out from conservatories or greenhouses, before planting out in borders during the warmer months, and returned to a safe shelter for winter.

Figure 1.4: Illustrating bedding flowers.

Source: https://www.pashleymanorgardens.com/the-gardens-2/history/pashley-manor-gardens-tulip-beds-lawn-and-lake-by-david-dixon/

Furthermore, the modern bedding flower industry nurtures and produces plants having a neat, dwarf trait, where the flowering is uniform and reliable. These are bred largely for utilization in large-scale bedding projects, where consistency and predictability play a paramount role. Even though this is usually achieved through losing the crops' individual character, leading some horticulturalists to term it an informal method of bedding (Thippaiah, 2005).

Forms of bedding: Typical bedding, as seen often in parks and big gardens, where entire flower-beds are replanted twice or thrice in a year, can be expensive and laborious. However, with proper management, it can still be achieved. Below are some popular flower bedding techniques used in towns and cities all across the world:

- Carpet bedding: Carpet bedding uses two or more complementary flower cultivars with a perfect, dwarf trait and unique coloring of flower and foliage, in order to develop geometric displays. It's often applied in the formation of things like or trademarks lettering, logos, floral clocks or even coats of arms. Suitable crops for this form of flower cultivation are rosette-developing succulents like Echeveria, or the rather slow-maturing or mat-forming foliage crops, including the tainted-leaved Alternanthera cultivars, that are tolerant of clipping. These plants may further be used in tri-dimensional sculptural formats or pseudo-topiary.
- Spring bedding: Crops used for spring-bedding usually are the biennials, which are sown one year for flowering in the next one, or hardy, though short-lived, perennials. Usually, spring-flowering bulbs like tulips are used in such situations, commonly with wallflowers, polyanthus and winter pansies.
- Summer bedding: Plants that are grown for summer bedding usually are annuals or the tender perennials. These become available in a form known as "plug plants." They can be found in flowering nurseries and garden centers especially in spring, where the crops can gradually be "hardened off" or adapted to outdoor environments by the buyer, before final planting around the period that the previous frosts are expected. Experienced flower growers keep a sharp eye on changing weather forecasts during various times of year. But while remaining on standby to safeguard their bedding displays during night hours using horticultural fleece, in case frost threatens.
- Winter bedding: Grown in autumn to provide a display that lasts up to early spring, most plants raised for winter bedding tend to be hardy perennials. They have to be planted during the same period of the year like spring bedding flowers, though winter bedding types are less commonly visible, except in containers like window boxes. A few are short-spanned and can be discarded immediately following their first display; whereas others can be applied as cutting sources for the successive year.

1.4.2 Perennials Flowers

A perennial flower, or simply perennial, refers to a plant thatlives for over two years. There are sources that further cite perennial flowers being those which live for well above three years. This term is widely used to distinguish a crop from shorter-spanned annuals and biennials.

Perennials, particularly small flowering plants, which grow and bloom during spring and summer seasons, usually die back in autumn and winter, before returning back in spring season emerging from their rootstock. These are further called herbaceous perennials. Nevertheless, based on the factors of local climate, a flower that's perennial within its native environment, or a much milder garden, could be considered by the grower as an annual which is planted out yearly, from the seed, cuttings or divisions of similar flora. Tomato vines, for instance, live multiple years within their native tropical/ subtropical environment but maybe grown in a similar way as annuals for temperate regions since they don't endure the winter (Anderson, 2015).

There's further a group of evergreens, or non-herbaceous, flower perennials, including plantings like Bergenia that retain a shroud of leaves all through the year. On the contrary, intermediate groups of plants are known as subshrubs, considering that they retain a vestigial wood-like structure in winter. Local climatic conditions may determine whether flowering plants are considered shrubs or perennials. Case in point, different categories of Fuchsia are classified as shrubs in warmer areas, but in colder temperatures may be chopped back to the ground each year due to winter frosts.

Lifespan:A Perennial plants may be short-lived, spanning just a few years, or long-lived comparable to some woody plants. These include a wide array of plant groups ranging from liverworts to ferns, including the widely diverse flowering plants such as orchids or grasses. Crops that flower or fruit just once then die may be classified as semelparous or monocarpic, even though most perennials can be categorized polycarpic or iteroparous, meaning they flower over multiple seasons throughout their lifetime. Perennials usually grow structures that allow them to acclimatize to persist from year to year through a type of vegetative-replication instead of seeding reproduction. Some of these structures include tubers, woody crowns, bulbs or even rhizomes. Additionally, they might have special crowns or stems which permit them to persist through periods of latency over cold or arid seasons throughout the year. Annuals generate seeds to perpetuate the species like a new generation when the planting season is appropriate, plus the seeds can survive through the cold or arid period to commence growth when the environmental conditions are suitable once more.

Most perennials have formed specialized features thatallow them to endure extreme climates and environmental factors. Additionally, some have evolved to survive through hot and desiccate conditions or freezing temperatures. Such plants usually invest most of their resources into survival adaptations, and usually don't flower or set seed up to after the first few years following growth. The majority of perennials also produce reasonably large seeds, which may have a great advantage over other smaller seeded flowers. Larger seedlings formed after germination are in a better position to compete in the environment for healthier flowers (Anderson, 2015).

1.4.3 Cut Flowers

Cut flowers are those which have been derived from the parent crop bearing them, they are typically detached from the crop for decorative use. Some typical uses of cut flowers are in garlands, vase displays and wreaths. Some growers harvest their private cut flowers from local gardens, though there's still a significant flower-industry for the cut flowers in many countries. The flora grown in different zones vary by climate, culture and other socio-geological factors within the environment. Oftentimes, plants are nurtured specifically for this purpose, usually in field and glasshouse growing conditions. However, it's still possible tocut flowers to be harvested straight from the wild (USDA, 2019).

Figure 1.5: Examples of cut flowers.

Source: http://www.flowermuse.com/long-lasting-flowers-every-month.html

Cultivation and Use: Cut flower farming is more intensive compared to other planting methods, typically on the foundation of greenhouse monocultures, plus may require large portions of toxic pesticides, deposits of which may often still be seen in some flowers, especially the imported varieties.

Cut flowers are mostly used in floristry, mostly for decoration within a home or building, where typically they are placed inside a vase. A variety of similar kinds of decorations may also be used, particularly in social events like weddings. They might even be decorated with extra foliage in some instances. Plus, the flora may be picked or plucked from the original plant, instead of being cut, something that can prevent damage on the leaf or stem. These flowers may be applied for wearing in the hair, or fitted in a button-hole. Bunches of flowers may also be set up for sprinkling, similarly to confetti. Whereas wreaths, bouquets and garlands are curated or value-added varieties of cut flowers (USDA, 2019).

Longevity:When flowers are taken out from the crop, they continue growing slowly, though have a reduced capability of in-taking the nutrients that are essential for their survival. For most countries, cut flowers come in the form of local crops due to their perishable status. In warmer areas, much of the flower will have a shelf-life of just one day.

Among some known varieties are marigold floras for temples and garlands, which are normally harvested before dawn, but discarded following use that particular day. However, the greater majority of cut-flowers are projected to last for several days with adequate care. This largely requires standing them inside water in shade, plus they might be treated in different ways to improve their lifespan. According to horticulturists, originally placing cut flowers within a sterilized vase may be important in extending the flower's life. Vases may be cleaned by a home dish detergent, or combination of bleach and water (USDA, 2019).

By applying these disinfectants, it ensures there areless bacteria propagating within the vase which could potentially make the crop to wilt or perish at a faster pace. Additionally, experts claim that slicing the flowers diagonally using a sharp-knife under flowing water ensures they can instantly take up fresh and unpolluted water for fresh-looking petals. Re-cutting the stems occasionally will further ensure there's a fresh surface by which the stems may take up water, something that allows the plants to persist even longer.

1.4.4 Potted Flowers

Potted flowers are often considered as the answer to tough-to-plant areas, or table garlands for outdoor events. Most flowers grow optimally in containers, where they can be cultivated to endure until the season's end. Utilization of potted flowers further makes it ideal to grow crops in places where they would naturally not be planted, or aren't appropriate to your environment. Additionally, potted flowers are portable and can easily be transferred from one section of the landscape over to another. For instance, you may start the flowering seeds indoors but later move the plants outdoors once the weather conditions permit.

Figure 1.6: Illustrating examples of potted flowers.

Source: https://balconygardenweb.com/best-blue-flowers-to-grow-in-containers/

Patio containers: Huge patio containers usually have a dramatic sight impact, plus add extra height to a venue when planted using trellised flowering climbers such as honeysuckle. As a grower, you might also prefer flowering shrubs to use for the patio container, like Azaleas which are also known as Rhododendron spp., or gardenias that are called Gardenia jasminoides as well. Other varieties like Rose-of-Sharon or Hibiscus syriacus particularly do well in big containers, but for optimal growth, you should feed your potted flowering plants while keeping them watered in order to enjoy their captivating scents and blooms. Likewise, pruning keeps the plants flowering and promotes growth in thedesired manner.

Window boxes:These are potted flowering plants thatfill up a window box, adding beauty to the interior as well as exterior of a home. As for window boxes found in shady zones, Impatiens would be more appropriate. In places that receive high amounts of sunlight, crops with sage-like Salvia spp. will be suitable, or other resilient sun-lovers. Annuals further perform well in window-boxes, like the French marigolds or Verbena. Both species do quite well in containers, plus are ideal for gracing your window-box with vibrant colors including gold, yellow and purple just to mention a few.

Herbs:Flowering herbs, like lavender which thrives mostly in temperate climates, and chamomile make outstanding container crops for the residential garden. You may keep the herbs restrained, with no need to worry about them spreading to adjacent flowerbeds. Additionally, potted flowering herbs particularly look impressive perched next to the window or kitchen door. These plants also fill up the air with their sweet scent as one passes by. Equally, Nasturtiums and Pansies are annuals that produce edible flowers that people can occasionally eat if they want to.

Hanging baskets: Potted flowers, which are compatible with hanging baskets, should feature an upright growing pattern that cascades steadily over the pot. Examples of these flowers are wave petunias, classified as tender perennials, and fuschia which is a frost-tender perennial plant. Just 2 or 3 of these flowering plants are needed to fill up a basket since the plants reproduce quickly in the growing period. If you select a blend of flowering crops for the hanging basket, check to ensure that each plant has equal growing requirements. Case in point, annuals like moss roses prefer a dried soil, though impatiens require a humid planting medium. In case you try keeping up with the watering requirements for impatiens plants, you may waterlog the entire moss rose.

1.4.5 Succulents Flowers

Succulent flowers, also commonly succulents, are crops with parts thatare dense, fleshy and engorged, which is usually an adaptation for retaining water in hot climate or soil environments. The term "succulent" is derived from Latin word sucus, which means juice or sap. They may keep water in different structures, like leaves or stems, while some plants are also known to do so in flowers.

Therefore, succulent flowers have the capacity to survive unconducive hot seasons by using moisture stored in their thick plant parts to sustain themselves. In horticultural practice, the phrase "succulent" is oftentimes applied in a way that excludes vegetation that some botanists consider as succulents, like cacti. These plants are typically grown as ornamental, due to their remarkable and exceptional appearance, including their capacity to thrive with rather minimal care (USDA, 2019).

Scientists estimate there are around 25 plant families thatcan be classified as succulent, with some common ones being Crassulaceae, Aizoaceae and Cactaceae succulents. The natural environments of these moisture preserving vegetation are usually in places with high temperatures or low rainfall, like deserts. Succulent flowers have the capacity to flourish on limited moisture sources, including mist and dew that makes them capable of surviving in an ecosystem containing scarce water sources.

Cultivation:Succulents are highly valued as home plants for their beauty and ease of maintenance. If well potted, they require very minimal care to survive indoors. Also, succulents are versatile or flexible house-plants that can flourish in various indoor environments. Nevertheless, for most growers, over-watering and related infections form the primary reason fordemise in succulents. These crops can be spread in different ways, with the most popular method being vegetative propagation. This includes clippings whereby several inches of leafed stems are cut, then after healing they produce a callus.

Following a week or there about, roots may germinate. A second technique is a division comprising of uprooting an over-matured clump and drawing the stems, including roots apart. Yet another process is propagation through leaf, by providing for the development of a callus. For this technique, a bottom leaf shall be fully detached from the vegetation usually by twisting or slicing. The leaf thereafter dries out then a callus forms inhibiting the leaf from absorbing too much humidity and thus rotting. Mostly, this flower planting method takes just a few weeks before delivering healthy roots which will eventually develop new plants. Plus, the vegetative propagation may be different depending on the particular flora species (USDA, 2019).

Natural habitats:Apart from Antarctica, succulents are found in all other continents. Though it's often believed that most succulents originate from dry zones like deserts, steppes and semi-desert, the globe's driest regions don't actually make for appropriate succulent habitats. For instance, Australia, the world's most arid continent, has very few local succulents due to its frequent and protracted droughts. Even Africa, which is regarded as the continent having the highest amount of native succulents, doesn't host most of these florae in its driest regions. Nonetheless, while succulents won't grow normally in some of the harshest conditions, they still are capable of growing in conditions that are unsuitable for other plants.

1.5 ECONOMIC IMPORTANCE OF FLOWER PRODUCTION

Flowers play a significant role in boosting the economy when produced commercially. Being versatile products, they are not only used for aesthetics and decoration but many other roles as well. Below are some of the economic benefits of flower production:

1.5.1 Perfume Making

Figure 1.7: Flowers are widely used in the commercial production of perfumes and fragrances.

Source: https://www.arenaflowers.co.in/blog/how-to-make-your-own-floral-perfume-natural-perfume-recipes-with-flowers/

Some perfumes are developed from scented flower oils, which comprise of a blend of special essential oils derived from steam distillation of crushed petal roses. Also, an associated blend is flower water, which is typically applied for cooking, makeups, medicine and even religious ceremonies. The production method originated from Persia, before spread through Asia and India, plus more lately into Eastern Europe (Earthcaregreenhouses, 2015).

Rose flower is among the most popular blends applied in perfume making. In Iran, Germany and Bulgaria, it's damask roses or 'Trigintipetala' that arethe majority used. In other different areas of the world, the Rosa and Centifolia blend is commonly used. This perfume oil is translucent pale yellow, or tan-grey in color. Additionally, 'Rose Absolute' is a solvent-extracted perfume oil, derived from hexane catalyst to generate a darker oil, which is typically deep yellow-to-orange in shade. As for weight, the oil extracted is roughly 1/3000 to 1/6000 of the overall mass of the flowers; case in point, around two thousand flowers are needed to create 1 gram of oil.

1.5.2 Beverages and Drink

Edible petals or flowering buds are occasionally applied to flavor regular tea or mixed together with other herbs for purposes of making herbal teas. For instance, in France, there's a upswing inthe use of flower syrup, which is commonly developed from a derivative of flower petals. Equally, in the Indian region of Rooh Afza, they serve a concentrated squash delicacy made from roses, also popular are flower-flavored frozen desserts like kulfi and ice cream.

There are flowers thatare also being taken as food, including flavoring or adding their aroma to food. Yet other minor applications include candied flower petals, as well as rose-creams, which are basically rose-flavored fondants blended in chocolate, or topped up with a crystalline rose petal. These confectioneries are a conventional English treat widely found across the UK. In addition, flower nectar is sometimes used to produce jelly, marmalade, jam, or even soup, it's highly valued for its high vitamin C ratio and nutritional value. Their petals might also be pressed and sieved to make special flower syrups (Earthcaregreenhouses, 2015).

1.5.3 Medicine

The flower R. canina is commonly used to produce vitamin supplements. The capsules of many flower species have substantial levels of vitamins, plus they've been used predominantly as food supplements in many cultures around the globe. In traditional Chinese medicine, some plant species have been applied for cure stomach problems, while others are being explored for managing cancer cases. In pre-modern medicine, a compound of rose flower called diarrhodon has been adopted to make medicinal and therapeutic products, where red roses make up the key ingredient.

Figure 1.8: Preparation of floral medicine.

Source: https://hospiceministries.org/2019/09/17/sunshine-food-and-medicine-for-the-soul/

1.5.4 Art and Style

Flowers have a long cultural significance, dating back to ancient civilizations like Greece, where they were commonly associated with supernatural deities. There are also a number of antiquity paintings that feature flowers in the background.

1.6 GLOBAL FLORICULTURE OUTLOOK

It's estimated that the global floriculture industry would hit $77.30 billion going into 2026, expanding at a rate of about 8.8% from 2018 up to 2026. Flower growing involves farming of ornamental plants, pot flowering plants and nurseries.Currently, the Netherlands is the axis around which the global floriculture business revolves. Ranking as the biggest exporter of flower products. Because of the high amounts of supply and need in the country and Europe at large, the costs of flower products are also set to increase.

The international floriculture industry has different active participants from all the continents. Plus, the number of players, in developed and still developing countries is growing continuously. Currently, worldwide exports and imports on floricultural items exceed US$9 billion, whereby Europe is regarded as the largest importer as well as exporter of floral products. Global trade in flowers is quite complex since there are various factors that determine exports. These aspects include the level of demands coming from diverse types of markets, usually set by state regimes and worldwide buyer activity, brief life cycles of some leading variations of floricultural products, consumer preference changes, and disparities in style and consumption (USDA, 2019).

Emerging dynamics: The international floriculture market trend is largely controlled by advancements in digitization and online-commerce, where it's now possible to purchase flowers online and have them delivered right to your doorstep no matter your location. Additionally, flowers are commonly used at opening ceremonials or large receptions in different areas of the world. Growing internet penetration is assisting in creating awareness for floral products among customers all over the world. Case in point, during earlier times a great portion of the cut flora found in Europe came from Dutch flower auctions. But in more recent years, IT structures have been brought into the auctioneering and marketing processes, which has had a considerable impact on trade. Digitization in the flower business includes the auction's distance ordering system, use of online stores by wholesalers, including other intermediary services for both auction and wholesale flower dealers.

Improvements in e-commerce are expected to open opportunities for completing direct sales together with buyers, rather than using intermediaries in the flower business chain. Furthermore, 90% of the global floriculture business is presently duty-free worldwide. Majority of the commerce occurs under Free-trade treaties or preferential trading systems with developing nations in South America and Africa. Still, rising numbers of flower growers in developing nations and the movement of bulk sellers to these countries poses a challenge for the flower market in Europe (USDA, 2019).

Floriculture Market Analysis:The International Floriculture Market is subdivided with the presence of both global and local players in the marketplace. To drive more consumer awareness, traders are increasingly using aggressive marketing approaches and high visibility products to generate demand for their different flowering products. Heightened rivalry among leading flower industry players is also influencing marketing strategies, including pricing dynamics. Producers are also increasingly focusing on improvement of their floriculture products, through enhancement of infrastructure used for storage and shipment to customers. Some key players in this market include; Oserian Group, Syngenta, Karuturi Global Limited, Beekenkamp Group, Verbeek Export B.V. and the Danziger Group among others (USDA, 2019).

1.7 CAREER PATHS IN FLOWER CROPS PRODUCTION

1.7.1 Research

Flower researchers or scientists conduct fieldwork and studies to determine the best ways for improving yield quality and quantity. Sometimes, they may perform experiments in flower farms or laboratories, then write up subsequent reports in offices. Most of them work for established research institutes, private industry, pharmaceutical firms or even the federal state. Travel may occasionally be required, plus fieldwork could involve spending several hours in greenhouses or open fields where the flowers are cultivated. Additionally, some researchers perform studies in breeding, composition, production, yield, as well as management of flowers during harvest time, storage and supply stages. They could also investigate the chemical, biological, physical, and mineralogical factors of soils in relation to flower growth and health. Even going further to classify or map out soils, and examine the effects of alternative cultivation practices on flora productivity (study.com, 2019).

Step for qualifying as a researcher: The first requirement for this career is earning a Bachelor's Degree in an agricultural field, such as biology, botany or ecology among other disciplines. Having a higher qualification instatistics or flower economics is also a huge plus. Also, it's required that students must participatein some on or off-campus internship programs. This shall help them attain the necessary research and work expertise in plant science and botany, particularly in the field of floriculture. Interns may spend weeks or months studying flower yield and reproduction, assessing plant growth, or restraining pests in order to create an effective and improved yield supply. Equally, certification and licensing areneeded before the researcher can be allowed to practice.

1.7.2 Education

An education career in flower growing may involve teaching students in a college setting, or training cultivators directly on the best practices to apply when planting their flowers. For commercial farms, they may further be charged with providing managers advice on daily planning, supervision, organization and controlling cultivation activities on planted flower estates. It's their duty to provide insight intothe best floriculture practices, including emerging technologies that will ensure planting operations run proficiently and smoothly.

As an educator, you may specialize in various areas of flower plant anatomy or physiology, including genetics and ecological factors. Based on the particular role, your training work could include giving skills on identifying, categorizing, recording and assessing flower plant species and biodiversity. This is particularly essential for growers who constantly seek to improve their flower breeds, by developing more advanced species and varieties.

Moreover, tutors may be hired for ecological consultancy work by large floriculture farms, where they may be responsible for conducting analyses and environmental impact valuations touching on the conditions used for growing flowers, including ways in which secondary pollution coming from pesticides and other chemicals applied to flowers can be minimized. For commercial farms, this may even extend to identifying and refining chemicals produced by flowering plants for application in other products like drugs, fabrics, solvents, food, and construction materials. While in academia it involves presenting study results in journals, publications or even academic conferences. Also training and overseeing junior staff members within the college and volunteers, or managing botanical flower collections found in the institution where they are hired to teach (study.com, 2019).

1.7.3 Production Management

This role involves ensuring that the flower products produced by a farm meet up to the established standards of quality. In larger flower companies, there could be close ties between production administration, and overall or strategic management including marketing or finance duties.

Otherwise, typical responsibilities for the job may include; planning or organizing production time schedules, evaluating project dynamics and resource requirement estimations. Whereby they will calculate how much resources should be put into a particular flower growing project, vis-a-vis the expected returns it would bring to the producers. Also on the same note, production managers negotiate and agree on budgets, including timescales of delivery with clients and directors within the flower business.

Equally, they ensure that safety and health standards are met during the entire production process, from the time flowers are picked from the farm to storage, packaging and delivery to the market. In some organizations, the managers are responsible for ordering and buying materials used for flower farming, organizing for repairs and continuous maintenance of production tools and apparatus, or even determining quality-control standards.

In terms of qualifications and training needed, there aren't many qualifications required in flower production management roles, though most managers possess at least a degree/HND in related fields such as agricultural, horticultural, floriculture or land management. Entry requirements for pursuing these courses may include an understanding of English, Mathematics and Science courses. Nevertheless, many courses also accept practical expertise in fieldwork instead of academic qualifications. Candidates must demonstrate the ability to work with minimum supervision and personally initiate work when it's needed (study.com, 2019).

1.7.4 Writing & Communication

A writing andcommunications specialist focuses on the fields of preparing written content, editing, branding, as well as media and public relations. Being part of the flower company's corporate communication unit, they also work closely with Managing Directors, Global Market Coordinators, and local Country Directors for purposes of improving internal and external communique for the sector.

The officer will support a brand-driven communications strategy which positions the business within the floriculture sector, as a preferred partner for conducting business with and strengthening the corporate brand by highlighting their achievements in the sector.

Responsibilities: With support from the flower production team and corporate communication unit, the writing and communication officer will:

- Evaluate the existing communications functions and provide suggestions for improvement.
- Create a clear communications policy for the company and provide clear objectives.
- Manage the company's flower website and guarantee that it's updated with regular news content, appropriate publications and

striking stories about their products and work in the community.

- Provide floricultural content to corporate journals like brochures, leaflets, annual reports, magazines and videos.
- Manage the generation of professional standard publications, as well as key communications materials like annual reports, brochures and newsletters. Plus where appropriate they can also use professional writers.
- Foster constructive working relationships with various media agencies, appropriate correspondents and journalists in the sector.
- Provide advocacy initiatives by managing relevant audience research, collecting and analyzing appropriate data, as well as preparing materials.
- Capitalize on floriculture conferences, workshops and shows for purposes of profiling the flower company in a good light within the public.

1.7.5 Sales and Marketing

A flower business sales & marketing expert sells, demonstrates, and focuses on promoting products related to the company. Some of the products they deal with may even include online brands since nowadays many flower companies have a presence on the internet as well. Thorough knowledge and understanding of the floriculture industry and sales also form essential components of this role. Generally, the education needed for this job involves a bachelor's qualification in a discipline like agribusiness.

Apart from contributing to sales, these professionals also provide customer service being representatives of their respective companies. Their roles may vary from simply selling flower products, to other related items like recipes, decoration ideas or edibles made from flower parts. Of most importance to employers is getting a sales representative having a strong background in the flower business, including specific knowledge of the sector.

Training:Some schools provide a certificate, associate's or bachelor's degree courses in floricultural sales, or agronomy with sales and marketing perspective. The skills attained in a bachelor's or associate's course in general marketing and sales, may also be adapted into flower sales. A typical program seeks to develop the learner's confidence in interacting with potential customers, comprehending their motivations and fulfilling their needs. The courses may cover a variety of topics like; corporate communications, services marketing, personal presentation, product marketing and sales potential management. Usually, flower sales programs usesimilar topics found in the agriculture department, but there are specialized colleges that only focus on the flower industry and not agronomy in general (study.com, 2019).

1.7.6 Parks and Gardens

Parks are public spots used for relaxation and unwinding by members of the public. They vary in size from a tiny neighborhood park to others as large as New York's central park. As a parks & gardens manager, your role will be overseeing the maintenance and daily operations of these recreational centers.

Parks don't just vary in scope, but also in the way they are managed or administered. In America, these experts are employed within all sectors of government where there are parks. The National Parks Services housed within the U.S. Interior Department oversees all national parks in the country.

Duties & Responsibilities: The park worker's main role is ensuring ornamental plants and flowers found around the yard are kept in good shape. In some cases, the professional may be required to have a bachelor's qualification in natural sciences, landscape architecture, leisure studies, or other similar fields. Plus candidates with any related experience may qualify for a parks manager role, even though having an unrelated bachelor's degree. As for the experience, park managers need to have some background experience working within the areas of landscape architecture or public parks. Supervisory expertise may also be beneficial, especially for park managers in high positions of influence, where they oversee other workers operating under them.

Apart from basic maintenance, park managers should also provide long-term vision covering the park's preservation as well as public enjoyment, including leadership and motivation for staff members, volunteers, associates, and the public in general. As for job prospects, parks & gardens management positions are expected to grow by 6% between the years 2016 to 2026 (study.com, 2019).

1.7.7 Owner and Entrepreneur

A florist store is a retail institution that sells ornamental plants and cut flowers. The trade covers activities like flower care, floral design, flower arranging and merchandising, usually also flower delivery. These floral shops are becoming increasingly popular and among the most sought after searches online in the horticultural business. It's widely believed that following diamonds, flowers make up the most beloved gifts for girls. Therefore, launching a flower business will most likely give you good returns for a long period to come.

While setting up a florist store is relatively easy, and your business can be up and rolling within no time. There are actually three crucial issues that must be addressed so that the florist shop can become successful. First, it's the business location, your preferred site should be within a high-traffic and visible zone, so as to improve walk-in business. The other factor is delivery. Since flowers are perishable products, your business should offer customers a quick and efficient delivery package. The third crucial factor of trade is marketing and exposure. The company must join a flower association thatpromotes professional and safety practices, including widespread delivery no matter the geographic location. Since some florist stores may gain up to 25 percent or more of their trade, coming from individuals who make orders from outside the florist's normal service area.

Starting up a floral business doesn't need much capital either, with a budget of only $10,000 - $50,000 you can open a storefront within your neighborhood in no time. Also, some of the accessories that will be required to maintain the flowers include vases, ribbons and pruning equipment (study.com, 2019).

2

MAJOR TYPES OF FLOWER CROPS

CONTENTS

2.1 Introduction 28

2.2 Bedding Flowers 29

2.3 Flowers 36

2.4 Potted Flowers 41

2.5 Indoor Foliage 46

2.6 Cut Flowers 52

2.1 INTRODUCTION

Flower crops are grown through a discipline called floriculture. It comprises the growing of flowers for gardens and for floristry. Flowers make up around 70-80% of all plant species in the world.

Flower crops undergo sexual reproduction to produce flowers. Flowers act as the plants' reproductive structures, having both male and female parts (hermaphroditism). Flowers are specialized shoots that are determinate, have an altered stem with internodes that are pressed together, have altered leaves with many functions and are often clustered in a larger bunch. For fertilization to occur, pollination has to occur first. There are two methods of pollination; self and crosspollination. In self-pollination, the pollen grains are transferred from the anther to the stigma of the same plant. In crosspollination, the pollen grains are transferred from a different plant. After the fusion of the male and female gametes, a zygote is formed. This later develops to form an embryo.

For a plant to flower, it requires a specific day and night length called a photoperiod. A change in day length causes a photoperiod mechanism in leaves, thus releasing a florigen hormone thatallows formation of flower buds thus flowering.

Flowers have economic importance.

Traditional flowers can be used for;

- Hair embellishment and garland making
- Decoration purposes
- Perfume extraction (Jasmine)
- Pigment extraction (Marigold and Gomphrena)
- For their medicinal value (Rose and Jasmine).

Modern flowers can be used for;

- Vase and bouquet preparation
- Making corsages (orchids)
- Landscape decoration and gardening
- Floral ornaments and exhibits and even for thought expression

Types of flower crops can be divided into;

- Bedding flowers
- Perennial flowers
- Potted flowers

- Indoor foliage
- Cut flowers

2.2 BEDDING FLOWERS

These flowers are relatively short and blossom for a long period of time. They are mainly used for display.

2.2.1 Petunia

Petunia is an annual herbaceous plant that is a result of the crossbreeding between Petunia axillaris and Petunia integrifolia. Petunia is recommended for container planting, general garden use, border edging, hanging baskets and mass planting.

Figure 2.1: These are pink petunia flowers.

Source: https://www.almanac.com/plant/petunias?amp

There are four categories of the Petunia variety.

- Grandifloras – These are large-flowered petunias accompanied by both single and double flowering configurations. It was developed in the 1950s 'Ballerina' as the first hybrid Grandiflora.
- Multifloras – These have smaller but more flowers than the grandifloras. There are both single and double-flowered

Multiflora petunias. Singles are mainly used for mass planting in flower beds, while double ismore suitable for containers. It was developed in the late 1940s with 'Comanche' as the first hybrid multiflora.

- Milliflora – These are compact and miniature and produce large numbers of 1 to 1 ½ inch diameter flowers. Their compact structure makes them suitable for containers and edging beds and borders. It was developed from a lucky genetic mutation from Petunia x hybrida, ensuing in a true miniature plant.
- Spreading petunia – These can spread for 2-4 feet. They are perfect for hanging baskets and as an annual ground cover. It has more of a tropical nature than other petunia types.

Breeding of Petunia flowers can bring many desirable effects such as; developing new flower colors, creating interesting and unusual patterns, modification of plant growth habits, developing dwarf petunia plants for ornamental purposes, creating petunia that can withstand a variety of conditions and evaluating petunias with excellent performance in the garden as a bedding plant (Singh, 2006).

2.2.2 Busy Lizzie (Sassy Little Flower)

Impatiens walleriana commonly called Busy Lizzie or Impatiens is an annual outdoor blossoming perennial that only grows in warm climates. They do not demand much and are easy to grow. They shine as indoor container plants as well as in borders, raised beds, window boxes and containers outdoors. They are of the plant family Balsaminaceae and originate from the East and South East of Africa. They are herbaceous, strong and have a bushy growth. They grow up to 20-40cm but can exceed and grow to a height of 70cm. They blossom in red, white, orange, rose, pink or violet colors. They mainly bloom from May to October and are sensitive to frost. They do not do well in freezing temperatures (Singh, 2006).

The name Busy Lizzie comes from the plant's ability to bloom the whole summer until fall starts without fail and thus making them busy the whole time.

Figure 2.2: This shows the high diversity in the colors of the Busy Lizzie.

Source: https://www.stjohnsgardencentre.co.uk/impatiens-bizzy-lizzy/

The Busy Lizzie is preferably grown in locations with a light shadow. Directly exposing them to the sun could burn the flowers' sensitive parts while putting them in completely dark areas affects their blooming. They should also not be exposed to excessive wind as their fragile and their optimum temperatures for growth are between 20-24 degrees Celsius. The flower can be grown in soil with moderate to no compaction, well-moisturized soil with good drainage. Compost can be added to the soil to improve nutrient availability and also make it more permeable.

There is a large variety of the Busy Lizzie flower. Some are;

- The Impatiens walleriana Victorian rose with dark red and dark pink color reminding one about the rose
- The Impatiens walleriana Blackberry ice characterized by its fuchsia color
- The Impatiens walleriana candy series with a large color variety and has large blossoms each with five blossom leaves
- The Impatiens walleriana Fiesta series characterized by its extensive riffled leaves
- Impatiens walleriana Elfin with unfilled blossoms that reflect light intensively

2.2.3 Geranium

These flowers were introduced in Europe by the Dutch traders from South Africa. They were mistakenly grouped into the same genus as the hardy wild geraniums that were already growing in Europe due to their resemblance. It was later proven that there was a difference due to their petal shapes, stamen number, etc. reclassifying them as Pelargonium. They are still called Geraniums.

Figure 2.3: Geraniums grow in clusters on one stem.

Source: https://diamondsgreenhouse.com/?page_id=42

They are easy to care for and blossom abundantly with brightly colored petals from spring to fall. They may grow 2-4 feet tall and wide. They mostly grow in full sun with partial shade. They should be protected in hot extreme climate especially in the afternoon. Despite the fact that they blossom from spring to fall, they can blossom in winter only in temperatures above 45-50 ^{0}C (Milio, etal . , 2017).

They bloom in many different colors; pink, white, purple, orange, mauve and even reddish black.

There are six different groups of geraniums;

- Zonal geraniums – The name is derived from a zone of darker color on the leaf which can be very visible. The flowers grow in single, semi-double or double clusters.
- Ivy geraniums – can trail and spread up to 48 inches. Their leaves are mid-green in color and are loved. They can bloom in red, pink, white or purple.
- Martha Washington geraniums – These have rounded partially toothed leaves that have a mid-green color. They can grow 12-48 inches.
- Scented-leaf geraniums – Their leaves produce a fragrance when their leaves are crushed. They have variegated mid-green leaves. They have a lemon-scented variation that can be used to fight off mosquitoes.
- Angel geraniums – These are bushier than the others. Their leaves could be crinkled and rounded. Sometimes they could have a scent. Their flowers are pink, purple or white.
- Unique geraniums – These are older and have a bad smell when crushed. Their leaves may be incised.

2.2.4 Lobelia

These are annually grown flowers. They mostly grow in blue but also grow in lilac, rose, scarlet red and other tropical colors. They can grow a maximum of 12 inches. They require full sun but they may require shade in extreme heat. They mostly blossom from spring to fall. They are grown in gardens with high pollination capacity such as havens for humming birds, rain gardens,etc. (Milio, etal., 2017).

Blue sea holly, Siberian larkspur, Lily of the Nile, Texas bluebonnets, Blue daisy and blue hydrangeas among others together with Lobelia are the 10% of the rare blue flowers in the plant kingdom.

Figure 2.4: Lobelias mostly come in lilac as seen above.

Source: https://www.amazon.com/Outsidepride-Carpet-Lobelia-Plant-Flower/dp/B004I7DJ7W

There are certain types that are commonly grown in the home garden;

- Edging lobelias – these are annual flowers grown in warm weather gardens. They are also tender perennials. These can have trailing or bushy habits. They can be grown as container plants. They have five petals that blossom in blue, white, purple or rose. Some petals may have white eyes.
- Cardinal flower – this is one of the most popular perennial variations they are known for their showy spikes of scarlet red flowers. This flower can grow in stream banks and marshes.
- Great blue lobelia – this is a wildflower that has blight blue colored petals. It is also called the blue cardinal flower.
- Lobelia speciosa – this is a garden hybrid of L. cardinalis and L. siphilitica that blossom for long summer period. The petals are pink, red, lavender or blue.

Lobelia inflata also is known as the Indian Tobacco plant have serious toxic effects, therefore are not commonly grown as garden plants. It has however been used for many years for medicinal functions.

2.2.5 Rudbeckia

These are also called Glorious Black-eyed Susans. They blossom in August. They are brightly colored with petals in shades of orange, lemon-yellow and gold. They can blossom for a long period of time with minimal care and also attract pollinators. It is a member of the sunflower family. They have cone-shaped heads, giving them their other name, Coneflowers. They flourish from late July to the first frost. They grow anywhere from 10 inches to 7 feet. Some cultivars display some red flushes and also bronze or mahogany. The center of the flower is brown but some might be green centers such as the Prairie sun.

Figure 2.5: The black eye with a cone-shaped head of the rudbeckia is their main determining factor.

Source: https://www.crocus.co.uk/plants/_/rudbeckia-fulgida-var-sullivantii-goldsturm/classid.3538/

Most species are short-lived perennials but they sow on their own often returning each spring from seed that aredispersed from the previous fall. Rudbeckia hirta and Rudbeckia fulgida are the species commonly grown in gardens. Like other wildflowers, they do not require much care and can tolerate drought and diseases. They can become overcrowded due to their self-feeding capabilities. If this happens, the extras should be removed and transplanted. While growing them, one should maintain adequate spacing to avoid water lingering. Water lingering can lead to powdery mildew. They should also be watered at soil level as well instead of watering them from above (Milio, DeMuro, and Martin, 2017).

2.3 FLOWERS

These are flowers that live for more than two years.

2.3.1 Phlox

This flower is popular due to its heady clouds of billowy blooms, its willowy stature and its no-fuss nature. It has a long flowering period from July to September. Its botanical name is Phlox paniculata. It is commonly called Garden Phlox, Summer Phlox, Tall Phlox and Border Phlox. It can grow 1-5 feet depending on the cultivar. It has five petals that are packed in dense clusters by the dozen. They range in size from 4-6 inches tall and 6-8 inches wide. It often has a sweet fragrant smell. Most have green leaves but some have variegated foliage with creamy white or yellow margins. Planting different varieties of Phlox, can give blooming flowers all summer and also autumn.

Figure 2.6: They have long slender stems with a cluster of flowers per stem.

Source: https://www.gardendesign.com/flowers/phlox.html

Phlox flowers have some advantages;

- Under good growing conditions, they can live for long periods of time, even lasting in the garden for decades
- When many other perennials start to fade in the summer, Phlox flowers retain their color
- They attract pollutants such as bees, butterflies and hummingbirds
- They hardly to rarely need staking

Some of the disadvantages are;

- They are easily damaged by powdery mildew
- They can easily be damaged from grazes by animals such as dogs and rabbits
- Although they often self-sow, they rarely retain the color of the parent plant.

Phlox flowers do not like dry spells and require water in droughts. Although one needs to be careful and water at the soil level to prevent powdery mildew. Improving airflow can also help prevent this fungus and also having adequate spacing between the plants (Milio, DeMuro, and Martin, 2017).

2.3.2 Bleeding Hearts

These flowers are heart -shaped and come inshades of pink, white or red. They hang intricately from arching stems. They blossom from late spring to early summer. Common bleeding heart plants die after flowering.

Figure 2.7: They have a characteristic appertaining as they hang on the length of the stem downwards. They have a unique heart-shaped flower.

Source: https://www.vpr.org/post/vermont-garden-journal-bleeding-hearts-bloom-our-spring-gardens

They grow up to three feet tall and wide. They prefer shaded areas but may be able to tolerate sun in certain zones. They cause toxic effects if ingested causing mild stomach effects. They can also irritate sensitive skin. Therefore, they need to be handled with care. Deers and rabbits do not associate themselves with this plant.

Bleeding hearts die back during summer due to the heat but they grow back during the spring. Once they are yellow and wilted, the plant can be cut back to the ground but one has to wait till then so as to ensure the leaves are done collecting and storing food for the next bloom (Milio, etal., 2017).

They should be grown in moist, fertile soil that has an alkaline pH. They should be watered regularly especially from spring until winter but one needs to consider the drainage of the soil to avoid water-logging.Bleeding hearts are resistant to many diseases and pests but can be susceptible to aphids. Pesticides can be used but one has to test the pesticide on a few leaves because they are sensitive to soap-based products.It is good to combine them with other shade-loving perennials that will replace the bleeding hearts when they die back, such as ferns.

2.3.3 Coneflower

These are easy to grow and blossom for months. They make good cut flowers and draw pollinators and birds. The petals come in yellow, orange, red and pink. They can grow two to five feet tall and 1 ½ to 2 feet wide. They prosper in full sun but some require shade in hotter southern climate to prevent them from burning. Most blossom from June to August or even later. The most common color of petals is purple but they also are grown in orange, pink, white, yellow, green and red colors. The purple species (Echinacea purpurea) is the most common in gardens but there are others like E. paradoxa, E. pallida and E. tennessenssis. They are also very good at attracting bees and butterflies and other pollinators such as honeybees and hummingbirds (Milio, DeMuro, and Martin, 2017).

They require at least six to eight hours of sunlight a day to avoid floppy stems. Sunlight can also prevent powdery mildew. They require regular watering of about an inch a week. A compost layer should also be added in the spring to provide nutrients or if the flowers are small and are developing poorly. The soil needs to be well-drained to prevent water-logging. They can tolerate drought but they do best in dry to medium moisture.

Figure 2.8: The center of the flower is shaped like a cone, giving them their name coneflower.

Source: https://www.gardendesign.com/flowers/coneflower.html

Echinacea is prone to a specific virus called aster yellow. It is caused by a Phytoplasma. Some of the symptoms to look out for are deformed flowers and yellow leaves with greenish veins. It is spread by sap-sucking insects such as leafhoppers. This virus cannot be cured therefore one has to dig up the plant and get rid of it to prevent it from spreading. Coneflowers are also susceptible to leaf miners, powdery mildew and Japanese beetles.

2.3.4 Columbine

These flowers are well known for their ability to attract hummingbirds, allowing birdwatchers to have a better view. It looks like a clover when young. They are drought-tolerant herbaceous perennials making them good for woodland gardens. They are classified as Aquilegia. They are in the buttercup family. They come in many colors some even have two colors. The blossoms can be pink, purple, blue, white, yellow, red or salmon. They generally grow two feet tall and wide. They blossom in late spring to early summer. They are grown in preferably a partial shade in well-drained soil. Their drought-resistant allows them to be perfect candidates for xeriscaping (Plasmeijer and Yanai, 2012).

The leaves of the Columbine plants often have a visible effect like that of leaf miners. The damage is not serious and some gardeners like it as it looks appealing due to its randomly variegated look. If one does not like this effect, they should inspect the leaves for larvae and crush them with their fingers.

Figure 2.9: Their leaves being purple can be mistaken for petals. They give a vibrant color that attracts the eye.

Source: https://plantcaretoday.com/columbine-flower.html

This flower is good for borders and beds because it blossoms in many colors, it grown in clumps and has relatively shallow roots. It is a self-sower and can live for long periods even three years and are easy to grow. They grow from seeds or plants.

2.3.5 Hostas

These are popular shade plants also called plantain lilies. They are easy to care for and bloom from spring to fall. They come in different colors, shapes and sizes. They have purple or white bell-like flowers that may sometimes release a fragrance. They grow in areas that are chilly to where frost is mild. Since they grow in many different sizes, it could be in thumbnail size to very big. They can be eight inches as miniature plants and even grow to as big as thirty to forty-eight inches. They prefer to grow in partial shade but can also tolerate heavy shade. Most hostas require sunlight and enough moisture (Plasmeijer and Yanai, 2012)

Figure 2.10: Their flowers may release a very fragrant smell.

Source: https://www.rhs.org.uk/advice/profile?pid=610

The moisture in the soil needs to be leveled and mulching helps to reduce fluctuations. Hostas cannot survive in dry conditions and the roots need to be encouraged to grow deeper by watering deeply.

Those exposed to excessive sunlight require more water. Slugs are prominent around hosta plants and this can be prevented by removing weeds and decaying leaves. When growing them in containers, the drainage needs to be good. This can be achieved by creating drainage holes and ensuring that the potting mix promotes drainage. They also need to be watered more than those that grow outside in the garden.

Hosta virus X can be spread through the transfer of sap from one plant to another with contaminated hands or equipment. It causes mottled leaf coloring, lumpy leaves or brown spots on leaves.

2.4 POTTED FLOWERS

2.4.1 Poinsettias

These flowers are from the native Mexican species Euphorbia pulcherrima which means 'the most beautiful Euphorbia'. They come in many different colors: deep crimson, peach, flaming orange and many other hues. Red is, however, the primary color. They flower in January and December making them the best Christmas decorations. They are also known as Christmas stars due to their star-shaped leaf bracts.

Figure 2.11: The bright colors of this flower allow it to look radiant indoors.

Source: https://www.capitalfarmcredit.com/news/poinsettias

The large colorful bracts are often confused for petals but they are leaves. At the center of each leaf bract are small yellow berry-like structures called cyathia which are the actual flowers. They are sub-tropical and can wither temperatures below ten degrees shortening their lifespan. In temperatures lower than ten degrees, they should be grown as indoor potted plants.

While indoors, they still require enough morning sun and they also need to be shaded in the afternoon or hotter parts of the day. They can grow and reach a height of sixteen feet. The soil needs to be well moisturized and dryness can be felt by touching the soil. Although they require a lot of water, they should be watered to a point of it running freely out of the pot's drainage holes and not more than that. Water-logging is not good for the plant. It may cause a lack of air and that results in damaging the roots. The potted plant should also be moved to a cool dry area especially at night to prevent root rot disease which is more common in temperatures below 60 degrees F (Singh, 2006).

2.4.2 Easter Lilies

It is a perennial bulb with white trumpet-shaped flowers that are large and have an amazing fragrance. It got its name from being a longtime Easter flower. Its taxonomic name is Lilium longiflorum. The flowers face outward from the stem. The stems can grow 23-36 inches long. It has narrow leaves that dark green and can grow five to eight inches long. The flowers can grow 5 to 7 inches long. When they are planted outdoors, they blossom in July and August.

Figure 2.12: They have a distinct trumpet-shaped flower.

Source: https://www.ftd.com/paraguay-pcg/easter-lily/paraguay/7b264429py/?country_id=PY

They require full sun to grow and can only grow in sandy loam soil that is rich in humus. It needs to grow ata pH of 5.5 – 6.5. They mostly grow white flowers. The bulbs grow in rich well-drained soils with medium moisture because Easter lilies have to remain moist. They can tolerate different weather conditions but prosper in a cool area for the first eight weeks to allow them to blossom. They don't prosper in a super warm climate but can withstand a wide range of humidity levels (Singh, 2006).

There are different varieties of Easter lilies;

- L. longiflorum white haven is a fully white Easter lily that can grow 2-5 feet high and have flowers that can grow to seven inches.
- L. longiflorumNelle White is the commercial flower.

- L. longiflorumDeliana has yellow flowers.
- L. longiflorumElegant Lady has pink fragrant flowers.
- L. longiflorum Trimphator has white flowers with a rose-pink center.

Easter lilies are not highly susceptible to pests and diseases but aphids can occur that can be rinsed off with water.

2.4.3 Shamrocks

These are of the Oxalis species. They provide winter foliage and flowers. The wood sorrel (Oxalis acetosella) and the four- leaf clover (Oxalis deppei) also have the name shamrocks. They produce purple or green leaves with either yellow, pink or white flowers. They are associated with St. Patrick's Day. They have three leaves that allegedly act as a sign of luck. Some people use clover and shamrocks interchangeably but they should not be used in that manner. All Shamrocks are clover but not all clovers are shamrocks.

They require direct sun, barely moist soils and cool temperatures. They fold their leaves at night and open up when the sun comes up. When the leaves die back, especially if the plant is potted, they need darkness to rest. In this period, they should only have limited watering and fertilizer should be withheld. This period can last a maximum of three months. Once the dormancy is broken, the plant should be moved to a sunny window or an area with bright light (Singh, 2006).

Figure 2.13: In many places, these are a sign of luck. They are associated with St. Patrick'sDay.

Source: https://www.pinterest.com/pin/515802963553589260/

During times of growth, the soil should remain lightly moist (watering 2-3 times a month). They grow from small bulbs that are planted in early spring. They require fertilizer on a monthly basis in the spring and summer when they are in their active growth phase. They prefer temperatures of 60-70 ^{0}C when the sun is up and 55-65 ^{0}C after the sun sets. The normal humidity in the house is appropriate for the plant. Shamrocks attract Aphids and spider mites which can be eradicated using 'green solution'.

2.4.4 Cyclamen

This is an herbaceous perennial plant of the family Primulaceae. They have upswept flowers and their leaves are well patterned. They grow from tubers. The flower petals can be in red, white or pink. They can grow up to eight inches tall.

Outside in nature, Cyclamen grows in humid and cool environments therefore, any temperatures above 20 ^{0}C during daytime and 10 ^{0}C at night will kill the plant, especially in the house. High temperatures yellow the plant and cause the flowers to fade.Cyclamen require proper watering. They can die if they are overwatered and also under- watered. The soil used needs to have proper drainage and should only be watered when the soil is dry to touch. They should be watered from the soil level to avoid the water from touching the leaves and stem. Dampening the stems and leaves can cause them to rot (Singh, 2006).

Figure 2.14: This clearly shows their upswept flowers and pattered leaves.

Source: http://www.burlake.com/4-cyclamen/

Cyclamen should only be fertilized once every one to two months only with water-soluble fertilizer that is at half its strength. If over fertilized, it can prevent the plant from blossoming. Cyclamen go through a dormant state that makes it look like the plant is dying due to yellowing and falling off of the leaves. They grow in slightly acidic soil.

There are different varieties of Cyclamen;

- Sierra series – they have large flowers that are about two to three inches that may be lilac, white, pink or purple.
- Victoria – these have white ruffled flowers with red margins. They undergo open pollination.
- Scentsation – these grow as either pink or red flowers with a strong fragrance.

2.5 INDOOR FOLIAGE

2.5.1 Bromeliads

These are lovely foliage plants that were initially believed to be hard to maintain indoors but can now be adapted to indoor conditions. They have red, purple, green, orange, spotted and even striped leaves. Its botanical name is Bromeliaceae. Bromeliads are epiphytes. They grow well indoors. Different varieties have different tolerance for sunlight. Some can stay under the direct tropical sun without any night effects but some would burn under such conditions. This does not stop them from requiring light to grow. They require well-lit areas such as bright windowsills but direct sunlight should be avoided. Yellowing can show that the plant is receiving too much sunlight.

Figure 2.15: They have unique long slender leaves. The flowers may be mistaken for leaves due to their orientation.

Source: https://www.mitre10.co.nz/gardenclub/article/bromeliads

Bromeliads can withstand drought conditions. They should be watered weekly as they grow but that should be reduced during their rest period in winter. They can tolerate temperature variations but they prefer temperatures of about 55 to 80 ^{0}F and cannot survive in temperatures below 40 ^{0}F (Singh, 2006).

There are different subfamilies of bromeliads;

- Guzmania – This is the commonest house plant variety. It blossoms flowers with yellow, red, orange, purple and white flowers.
- Neoregelia – These grow in bold orange, pink or red flowers.
- Pineapple – These have spidery leaves and miniature pineapples on the flower spike.
- Vriesea – These have feather-like blossoms and variegated foliage.

2.5.2 Ferns

This is one of the most modest plants. It remains at the bottom of the forest floor and resurfaces once in a while to unfurl its foliage. It procreates by sending dust-like spores in the air which land and grow into a tiny plant called prothallus.

Ferns do not produce seeds nor do they produce flowers. They reproduce using spores. Although spores are not the only way for reproduction to occur. In some ferns, it occurs through the branching of the underground portion of the plant. This part is called the rhizome (Singh, 2006).

Figure 2.16: Ferns do not produce flowers. They reproduce using spores.

Source: https://www.burncoose.co.uk/site/plants.cfm?pl_id=5326

Ferns can grow to a range of 1-1.2 cm to 10 to 25 meters in height. Some are vine-like and others float on pond surfaces. They mostly grow in damp warm areas. They are rarely found in cold dry areas.

Pteridium is one of the best-known fern genera. It is mainly found in forests and cleared land or old fields. They grow in the tropics mostly seen in tropical rainforests especially as epiphytes on trunks and branches of the trees. Ferns have not shown to be on economic significance to humans, but some have been used for medicinal purposes. Osmunda and Pteridium have been consumed in East Asia and have proven to be highly carcinogenic (Singh, 2006). The main economic significance of ferns has been in horticulture. There is a huge supply of both indoor decorations and outdoor gardens. Bracken which is poisonous often spoils the grazing value of the land. In many countries it is considered a weed.

2.5.3 Philodendrons

This genus has the most beautiful foliage plants in the kingdom. They have glossy leaves which remind people of the tropical areas such as the jungle. There are two types: the climbing varieties and the self-heading types. Those that climb are mostly used in hanging baskets and as nice indoor foliage plants in pots. These plants can grow extensively in the wild but are tame indoors.

Philodendrons originate in tropical areas. P. scandens are dependable house plants. They survive best indoors of all the other Philodendrons. They require a light that is similar to that found under a tropical canopy. They thrive in light shade. Yellow leaves are an indicator of too much direct light.

Figure 2.17: Their leaves have yellowish streaks that are pleasant and decorative.

Source: https://www.plantopedia.com/philodendron/

They require soil that is rich in inorganic matter and has good drainage. It needs to be loose as well. The soil needs to be moist at all times. Any roots above the soil for the climbing variety need to be pushed into the soil. Overwatering should be avoided to prevent root rot. Drooping leaves are an indicator of either too much or too little water. Philodendrons cannot grow in temperatures below 55 degrees F. They grow better in humid areas which can be controlled indoors using a pebble tray of water. Misting them also helps. Philodendrons produce larger leaves and are healthier when they are well fertilized. Slow-release pellets can be used at the beginning of the growing season. They can be fertilized once a month during the winter (Singh, 2006).

2.5.4 Pothos

This is one of the easiest plants to grow indoors. It is a vine with green leaves that are shaped like hearts and maybe variegated with white, yellow or pale green. They like bright lights but can also survive in areas without sunlight making then good for offices or even dorms. One of their greatest advantages is that they purify air indoors of chemicals such as benzene and formaldehyde. Pothos isnot able to support themselves on their own but they can be trained onto supports to give a twining appearance and also secured on hooks. Indoors they can grow up to 30 feet long. They are mostly pest-free but mealybugs can be a stubborn pest. One can use insecticide soap or alcohol-soaked swabs (Singh, 2006).

Figure 2.18: The leaves are variegated with white and are heart-shaped. They could also have yellow or pale green variegations.

Source: https://m.springhillnursery.com/product/variegated-pothos

Outdoors, they can grow in shade to partial shade. Indoors, they grow in bright indirect light. Without enough light, the variegated leaves turn all green. When the plants get too much sun, they turn to a pale color. They thrive inwell -draining soil with a slightly acidic pH. Pothos need to be left unwatered between waterings to avoid root rot due to damp soil. Signs that the soil is too wet are presence of black spots and collapsing of the plant. When dry brown edges occur, that is a sign that the plant is a bit too dry.

Pothos should be kept above 50 degrees F and in high humidity but it can also survive in low humidity. The stems should be trimmed to maintain foliage (Singh, 2006).

Some of the recommended cultivars are:

- Marble queen – these have white and green variegated leaves and need more light than other Pathos.
- Neon – these are bright and great for darker areas.
- Silver Satin – these have grayish greenish leaves that are thick and have silver splotches.

2.5.5 Devil's Ivy

This is a climbing house plant. Its name originates from its survival skills. It is a very hard plant to kill off. It is called Epipremnun aureum. It is originally from the French Polynesia islands. It is highly invasive and referred to as an uncontrollable pest. Its blooms have a sheath around them. The foliage has leaves that are heart-shaped and variegated. Epipremnun aureum is the most popular and the easiest to grow. The leaves can either be mid-green or mustard yellow.

Figure 2.19: These can grow on trees and into canopies in the wild.

Source: https://www.touchwoodflowers.com.au/product/devils-ivy-indoor-gift-plant/

Devil's Ivy can grow up to 400 meters and the leaves can grow up to 100cm in length and 45cm wide. It grows aerial roots that can grow and attach to any structure next to them allowing them to climb trees to canopies. Indoors, they can grow up to three meters if they have structures to grow on e.g., a moss pole.

Devil's Ivy grows well in temperatures of 18 to 24 ^{0} C. Devil's Ivy can grow with less effort in brighter lights and dark it will grow very slowly. The plant should be watered only if necessary with a diluted high-nitrogen foliage feed. These plants are beneficial as they purify the air from harmful household pollutants (Singh, 2006).

2.5.6 Wandering Jew plant

This plant is commes under tradescantia genus. It is an herbaceous perennial plant. The name Wandering Jew comes from the plant's habit to move to wet and moist regions. It has a flower with three petals that come in white, purple or pink and are not very conspicuous. It can grow to a height of six to nine inches.

Figure 2.20: Wandering Jew plant.

Source: https://www.etsy.com/listing/294761993/tradescantia-purple-x-5-cuttings

The most difficult aspect of growing a Wandering Jew is maintenance of the moisture. It grows well in moist soils It requires well-drain soil to prevent root rot that may occur if the roots are kept too wet. The plant does well in bright but indirect sun. The leaf variegation can fade when the plant is not exposed to enough light and too much light can burn the leaves. The

Wandering Jew can be planted in normal potting soil as long as it drains well and is not over watered (moist but not soggy).

The wandering Jew does best in temperatures between 65 – 75 ^{0} F but it can also thrive in warmer temperatures. Any temperatures below 50 ^{0} F can discolor and kill the leaves. The soil should be fertilized bimonthly. Over-fertilizing can cause the leaves to lose their variegation.

There are different varieties of the wandering Jew;

- T. pallida 'purple heart' – these have solid purple foliage.
- T. zebrina pendula – these are colorful and have reddish leaves.
- T. callisia – these have white stripes.

2.6 CUT FLOWERS

2.6.1 Roses

Roses come in many forms. There are climbing roses and even miniature roses. They blossom at the beginning of summer and during the fall. Roses are grouped according to the date they were introduced:

- Old roses – These were introduced before the year 1867. They were fragrant and are commonly seen in old paintings.
- Modern hybrid roses – these were introduced after 1867. These were bred for their shape, size, color and fragrant smell. They blossomed for long periods of time and they were resistant to diseases.
- Wild roses – these have been growing for many years. They have now been assimilated into modern gardens and blossom from spring to the beginning of summer.

Roses require a minimum of five to six hours of full sun a day. The morning sun dries the leaves up from the dew and is important in preventing diseases. The more sun the rose bush gets; the more flowers it will produce. Roses require well-draining soil but one that can hold enough moisture for the roots to absorb. Roses do not like soil that retains too much water. They don't grow well in soggy soil. They prefer loam soil that is actually leaning to the sandy side. They should be planted with space between then for proper airflow. This also prevents powdery and downy mildew. The soil needs to be at a pH of 5.5 to 7 (slightly acidic to neutral) (Singh,2006).

Figure 2.21: These are bred for their shape, size, color and fragrant smell. They are the most common cut flower type.

Source: https://www.jparkers.co.uk/rose-standard-fragrant-cloud-0001745c

In dry summer weather, the roses should be watered at soil level at least twice a week. The entire root zone should be soaked to avoid the growth of fungus which occurs in shallow sprinkling. Mulching can also be used to conserve the water and moisture in the soil which will encourage the plant to thrive. One should use compost and natural fertilizers before and during the blossoming cycle. Using artificial fertilizer can attract aphids. Removing dead leaves and canes also helps reduce pests.

2.6.2 Baby's Breath

These belong to the Caryophyllaceae family which also includes Carnations. Baby's breath is in the genus gypsophila. The flowers are tiny but they compensate for that in their volume. They can produce hundreds of flowers per plant. They can grow up to four feet tall with very slender stems. They can be grown as perennial flowers due to their hardiness.

Baby's breath grows well in full sun but it can also tolerate some shade especially in the afternoon. Too much shade causes leggy plants. They prefer well-draining soils (soils closer to sandy soil) compared to clay soil to prevent water-logging. The soil needs to be at a pH of 7 – 7.5. Lime can help increase the pH of the soil if it's too acidic (Singh, 2006).

Figure 2.22: The flowers are tiny but they compensate for that in their volume. They can produce hundreds of flowers per plant.

Source: https://flowergiftkorea.com/product/large-babys-breath-bouquet-1/

Some of its varieties are;

- Gypsophila elegans – this is an annual self-feeding species that can live a long time in the garden. It performs well in meadows.
- Gypsophila paniculata 'Bristol fairy' – these are white and semi-double.
- Gypsophila paniculata 'Compacta Plena' – these are a smaller variety that grows to one foot and do not require support.
- Gypsophila repens – these come in pink or white and are a creeping form.

2.6.3 Carnations

There are different types of carnations but the most popular are the border carnations, annual carnations and the perpetual-flowering carnations. Their scientific name is Dianthus. They are from the family Caryophillaceae and the genus Dianthus. They come in many different colors each with a meaning, making them very popular. A blossom has five petals and the petals can be white, pink or purple. They grow to about six to eight centimeters in diameter. Their petals are serrated. The flowers are bisexual and blossom in a forked cluster. The leaves are narrow and vary in color from green, greyish bluish or even purple.

Many carnations bloom in the late spring and can continue well into autumn. They should be grown in well-drained soil with a pH of 6.7 to 6.9. They should not be over-fertilized because that leads to very green leaves but very few blossoms. Wet soil that has poor drainage causes yellowing and discoloration of the leaves and that can even kill the plant. There should be enough space between the flowers for air circulation to prevent growth of mold and mildew. Pesticides are not necessary because not many pests affect carnations.

Figure 2.23: These are large-flowered carnations. They have one large flower per stem. They also come in many different colors and some even bicolor.

Source: https://www.flipkart.com/fourwalls-artificial-single-carnation-stems-set-15-red-pink-mix-red-carnations-flower/p/itmehrh2gekgdcd7?pid=ARFEHRH2GNG2DHG2

There are three types:

- Large-flowered – these have one large flower on each stem.
- Mini carnations – These have multiple small flowers.
- Dwarf flowered carnations – multiple small flowers on each stem.

They need a few hours of full sun on a daily basis and also moist soil. Overwatering, however, can cause yellowing of the foliage. Any withered flowers should be removed to promote blossoming of new flowers. Carnations can bloom for years if properly cared for.

2.6.4 Lilies

Their botanical name is Lilium. They are popular in summer gardens with a fragrance and a strong stem and act as workhorses as cut flowers. Their bulbs can be planted during spring but they grow better if they are planted during autumn. They should be planted on a raised bed to ensure proper drainage of the soil to prevent root rot and death of the plant. They should be planted in threes. They require rich soil but not overly fertilized.

Figure 2.24: Bicolor lily with a white border and dark spots. Lilies come in many shapes, colors and sizes.

Source: https://www.almanac.com/plant/lilies?amp

Mulching can help keep the soil moist and cool. Specially long -lasting mulch like wood chips or cocoa shells. Any spent blooms should be removed to promote more blooms. They prefer light partial shades. Some can even survive in rooms without sunlight. Yellowing leaves can show that the plants are getting too much sunlight. They prefer temperatures of 65 – 85 ^{0}F. The indoor lily is immune to insects and diseases. They can get aphids but these can easily be wiped off the leaf.

Some of the lily varieties are;

- 'Stargazer' lily – these grow up to three feet. It is an oriental hybrid lily.
- 'Casa Blanca' lily – these are white and fragrant with waxy petals. They bloom from mid to late summer and grow three to four feet tall.
- Turk's cap lily – these have pagoda-shaped lilies. They are the most shade tolerant lilies and more sensitive to heat.
- 'Tiny Bee' lily – these are small hybrids of the dwarf Asiatic lily. These grow best in windy areas where tall lilies may break.

2.6.5 Alstroemerias

This comes from the family Liliaceae. It is a perennial flower that grows twelve to thirty-two inches. They flower from June to September. They are very similar to lilies having trumpet-shaped flowers. They grow from tuberous roots. They grow best in locations with full morning sun and some afternoon shade. The soils need to be fertile and well-drained to prevent root rot and death of the plant. Most Alstroemerias are winter hardy. Their flowers appear at the top of the stem in clusters with colors ranging from white, red, orange, pink, mauve and yellow. One petal can have one color or two (Singh, 2006).

Figure 2.25: They are very similar to lilies having trumpet-shaped flowers. They, however, grow from tuberous roots.

Source: https://bloomingmore.com/products/alstroemeria-assorted-colors

They can be planted during spring or fall sometimes even the early summer. They should be planted when the temperatures do not exceed 68 degrees F. They require regular watering but not to make the soil soggy. They should be fertilized when in bloom at least once a week. Mulching using leaves or pine needles can help retain the soil's moisture. They spread in a bushy manner.

Some of its popular varieties are:

- Alstroemeria aurantiaca – these come in orange, yellow or red with dark red streaks and can grow as tall as two to three feet.
- Alstroemeria hookeri – these come in pink or apricot with yellow and brown marks. They can grow up to four to six inches tall.
- Alstroemeria pelegrina – these come in magenta pinkish with a white and yellow throat and can grow 1-2 feet tall.
- Alstroemeria ligtu – these come in pale pink, red or white color and grow 2-3 feet tall.

2.6.6 Chrysanthemums

These are popular during fall. They grow 1 to 3 feet depending on the variety. Their blossoms are of many different colors such as yellow, orange, white, lavender, or red. They can also be bicolored.They require full sunlight during the bloom period. Lack of enough sunlight will cause the plant to grow weak, producing fewerflowers. After the flowers develop, the plant could be moved to a shaded area for the flower heads to last longer. They prefer rich well-drained soils like that for vegetables. They bloom from September to frost.

Figure 2.26: Mums have many petals on one flower giving them their distinct cotton ball look.

Source: https://www.mr-fothergills.co.uk/Flower-Plants/Chrysanthemum-Plants/Chrysanthemum-Outdoor-Bloom-Plant-Collection.html

Mums are planted during spring, allows them to survive during winter. They have a shallow root system thus requiring more watering especially in areas with hotter climate and minimal rainfall. Mulching can help keep the soil's moisture, especially during summer. Mums don't needmuch nutrition, therefore they should not be over-fertilized. They are prone to some diseases like leaf spot and powdery mildew. Providing spaces between the plants can prevent this due to adequate airflow and it also prevents the plants from retaining moisture which can act as a habitat for some pests including aphids (Singh, 2006).

3

TASKS INVOLVED IN FLOWER CROPS CULTIVATION

CONTENTS

3.1 Introduction 62

3.2 Deciding What Flowers To Grow And Organizing Their Planting 62

3.3 Preparing Soil For Planting Flowers 64

3.4 Maintaining Flowers Which May Involve Spraying, Pruning, Fertilizing and Watering 66

3.5 Grading And Packing Flowers For Sale 68

3.6 Marketing And Selling Flowers 70

3.7 Transporting Flowers To Buyers or Exporters 73

3.8 Maintaining And Repairing Machinery Such as Pumps and Other Equipment 75

3.9 Hiring and Training Workers, and Ensuring That Their Work Is of a High Standard In Flower Farms 77

3.10 Ensuring That Health And Safety, And Other Regulations, are Complied With In Flower Farms 79

3.11 Ensuring Market Certification Systems (Such As Quarantine Requirements) Are Up To Date 81

3.12 Keeping Production And Financial Records And Write Reports 83

3.13 Researching Growing And Harvesting Techniques, and New Varieties 85

3.1 INTRODUCTION

Flower crops cultivation involves several processes. There are certain tasks that should be considered. Some of these tasks have been discussed in this chapter below.

3.2 DECIDING WHAT FLOWERS TO GROW AND ORGANIZING THEIR PLANTING

Flower growers need to choose the correct plants for cultivate so that they can grow to full maturity. Among the top errors that beginners make is picking the wrong flowering crop, which may eventually leave them discouraged if their yards don't bloom as expected.

Figure 3.1: Flower growers should choose the best breed to cultivate based on factors like soil pH, growth rate and ease of maintenance.

Source: http://bangladeshdiscover.blogspot.com/2011/06/floriculture-of-bangladesh.html

There are different types of flowers to consider:

Annuals: From initial budding to blooming, annuals persist for only one season. Annuals are valuable to any garden or person who prefers to get innovative from one year to another. These kinds of flora are usually cheaper compared to their perennial equals, plus will bloom throughout the season, meaning you have plenty of time to relish them. Some yearlings are also self-seeding, meaning you might end up with just a couple of the same florae the subsequent year. This is a key detail to remember in case you plan on planting new yearling flowers every 12 months. Some beautiful annuals for

3.3.1 Assessing Soil Health

While there are more than 17 compounds thatare essential for flower growth, phosphorus, nitrogen and potassium make up the most important nutrients. These are regarded as primary or macro-nutrients since plants uptake them from the earth in large amounts. When preparing your soil, ensure that fertilizers used contain all these three nutrients which are crucial for plant health. But they are barely complete in the absolute sense, instead calcium, sulfur and magnesium which are considered secondary nutrients, also play a significant role in healthy plant growth. Including lesser micronutrients such as cobaltthat is mostly utilized by plants as a fixer of nitrogen. Another crucial component for healthy soil is the acid-alkaline quotient or pH reading. When all these factors are balanced, including proper texture, it typically makes for a healthy flower planting soil (Wallin, 2011).

3.3.2 Testing for Nutrients

Among the effective ways of determining what nutrients are deficient or plentiful in your garden soil is getting it tested. There are local soil improvement experts who can provide affordable soil tests for you. These procedures typically measure amounts of soil pH, phosphorus, magnesium, potassium and even calcium, sometimes also nitrogen. Equally, they may record the soil's micronutrient ratio, but this isn't so much essential to gardeners who add ample amounts of organic material to their soil. For less intensive tests, there are do-it-yourself soil kits that allow you to do your personal simple and rewarding chemistry. Most of these kits have a 'color mode' for measuring pH, which turns to a different shade with variations in the soil's alkalinity or acidic levels. These ratios are critical for your flower's capacity to absorb minerals and nutrients from the soil.

Mostly, nutrients and minerals are most accessible to florae in soils having a pH amount of between 6.5-to6.8. In case soil conditions are acidic, with pH levels below 6.0, or alkaline where the levels are above 7.0, it won't matter how sufficient the nutrients may in soil , since the flowers won't be in a position to absorb them. The best period to get soil tested for acidity is during spring or fall seasonswhen it's the most stable. It's also an ideal time to do soil improvements, or apply organic fertilizer just in case garden is short of essential minerals or nutrients (Wallin, 2011).

3.3.3 Soil texture and Form

Soil texture is dependent on the quantities of sand, clay and silt that it holds. Generally, sand constitutes the largest pieces of soil specks and feels gritty when touched. It's followed in size by silt particles that are slippery while wet but powdery in dry condition. The tiniest fragments are clay.

These are flat shaped and usually, pile together like some plates or paper sheets, notrequire an expert to decide soil texture. Simply pick up some particles and rub in-between fingers. In case the soil is gritty, then that's considered sandy texture. But if smooth like talcum dust, it's silty. Additionally, if the particles feel harsh while dry and slimy and sticky in wet conditions, then soil is classified as heavy clay. The majority of soils fall somewhere in-between these textures (Vaughan, 2019).

Likewise, sandy soils often are nutrient-poor considering that water and nutrients quickly drain through their big spaces between the sand particles. These soils are furthermore low invaluable microbes and organic content that florae can thrive on. Silt soils, on the other hand, don't drain as much thus making them more fertile for flower germination.

3.4 MAINTAINING FLOWERS WHICH MAY INVOLVE SPRAYING, PRUNING, FERTILIZING AND WATERING

Flowers bloom when the correct growth conditions are set up for them.

Figure 3.3: Applying compost or granular organic fertilizers will feed your soil and plants with nutrients, which will lead to healthy plants.

Source: https://www.treloarroses.com.au/Fertilizing-Roses

Ideally, they should be applied at the start of the planting season when you're preparing the garden, including your containers. Also, for best results, spread them again during the close of the season in order to substitute lost nutrients or organic matter. That way, when spring arrives, your soil shall have plenty of residues to supply your flower garden. Other essential maintenance tips you can try include:

- Prevent weeds from growing near the flower. Clear up any debris from the soil, such as dropped leaves, twigs or gravel. Apply a 1-to2-inch top layer of mulch surrounding the plants using a garden fork or trowel. Ensure the mulch is 2 to 3 inches distance away from the crops' stems in order to prevent rotting.
- Water the plant only when the adjacent soil becomes totally dry. Add water from a typical garden hose directly onto the soil which surrounds the plants. Don't splash the flowers' foliage and wet the earth to depths of around 6 inches. Additionally, give the flowers water only in the morning to permit sufficient time for droplets of moisture to get dried up before nightfall.
- Remember to fertilize your plants at least once per month, using a 10-10-10 nitrogen, potassium and phosphorous water-soluble fertilizer. Blend 1 teaspoon of fertilizer together with 1-gallon of water, or depending on the instructions found on the package cover, then spread the fertilizer substituting it in place of watering.
- Cut off the plant's terminal end at the main stem for improved lateral branching in case the crop becomes tall and gangling. Place the cut around 1/8 to 1/4 inch just above the second set of leaves with a set of pruning snips.

Inspect the crop's leaves for possible aphids or mites, especially after watering the crop. Wash off minor insect populations using a steady jet of water during the morning. Equally, spray heavily infested floras with insecticidal detergent or horticultural oil for eradicating the pests (Vaughan, 2019).

3.4.1 Pruning for Specific Types of Plants

Shrubs and ornamental plants: The easiest method to trim flower bushes or shrubs is following the 50/50 rule of thumb. This means cutting off 50% of the plant's height and another 50% of the residual canes at the bottom section. However, this rule may sometimes need amending depending on the shrub's variety. Many hybrid plants produce slightly lesser but larger blooms compared to regular shrubs. To encourage high-quality flowers, you can prune their parent plants slightly 'harder' to provide room for fresh budding. Likewise, when pruning the canes, reduce their height by cutting them at side-angle, so that water may flow off the cut. It prevents the disease from developing on the ends. Climbing flowers: One common mistake that people make is reducing the height of their plant each year. However, experts recommend not pruning climbing flowers during the first 4 or 5 years. You can simply tidy them though don't start any significant pruning until after this period is over. When the correct time comes, simply cut 1/3 of the canes starting from the base, though don't worry a lot about decreasing the height, unless the plants are completely out of control. Once the plant's framework is established, the next step is slicing the laterals back in order to form short spurs which will then blossom beautifully the next season.

3.4.2 Pest Control

When it comes to managing pests, winter is the ideal time for combating fungal pests, including other opportunistic diseases. To begin, collect and discard off dead or parched leaves, and pruned branches. Thereafter, spray the plants with pesticide- including the ground surrounding them - with a mild organic lime-sulfur solution. Always ensure that pesticide isn't too much concentrated since this can burn the germinating plants.

3.5 GRADING AND PACKING FLOWERS FOR SALE

Figure 3.4: Flower grading and packaging is a delicate process that requires knowledge and accuracy.

Source: http://ghsasia.info/products/flower-processing-line/

Many farmers and horticulturists grow cut flowers that are sold in retail garden centers, farmers' markets, roadside stands, and even farm shares. Appropriate harvesting and handling of flowers following harvest arevital in maximizing the vase life, which ensures high-quality products for sale.

3.5.1 Harvesting and Grading

Flowers need to be harvested during the morning, soon after precipitation has dried up or in the evening, but never during the day's heat. In the early hours, temperatures are down and humidity content is also high. Make defined cuts to the crop to obtain long, solid stems. Take out the foliage found on the stems, which shall be below the moisture which will otherwise decompose and promote bacterial growth (Vaughan, 2019). Some flowers like sunflowers profit from the deduction of majority or their entire foliage. Besides, for some plants, side shoots found on the primary stem are usually removed during harvest time. Stems may be sliced on a slant or square line. Typically, the slant cuts will prevent stems from staying flat on the container's bottom and improve water uptake.

Flowers are usually graded and bunched soon after harvest and prior to placing them inside water. This practice lessens handling steps and reduces mechanical damages, which often happens on flowers and leaves. In case grading and bunching can't be completed immediately, then the pickings should be placed inside clean buckets, filled with clean tepid water and some biocide. The tepid, acidic water shall help reduce air bubbles. Additionally, avoid over-filling the collection buckets with flowers, in order to prevent unnecessary bruising and tangling. The water's depth inside the bucket should be considerably deep to cover the bases of the stems, ideally, 1-6 inches based on the stem's size and buckets available.

During harvest time, some gardeners put buckets at the very end of the lines where they put flowers as they chop. Flowers must never be laid directly on the bare earth, or dirty and soiled surfaces where they could collect dirt and pollute the stems or buckets. Florae and foliage stems must always be preserved clean. Apart from clean buckets, the cutting tools also need to be frequently disinfected. There are growers who place a bucket having a disinfectant solution, including clippers at the back of the row. This encourages gatherers to effortlessly exchange used equipment for clean ones upon finishing a row (Vaughan, 2019).

3.5.2 Packaging of Flowers

Proper packaging is important to preserve the condition of your flowers. Make sure your deliveries comply with all applicable local, statewide and federal laws determining shipment of floral products and plants. Also, the processes of preparing flowers or ornamental plants for delivery, right from floral arrangements to wholesale flowers, seedlings to matured plants, may vary depending on the contents found on your shipment.

For starters, secure floral arrangements tightly within a box using secure cable ties or bands fastened to the box. Alternatively, they can be fixed inside in order to prevent jostling or misalignment of the floral arrangement during shipment. In case you're shipping a vase that has an arrangement, it's also recommended to secure the vase within the box using a corrugated divider (Henderson, 2008).

When shipping ornamental plants, secure them well to prevent movement within the shipping carton. In case the plant swings inside the carton, this can cause breaking of plant stems including the leaves, so it's advisable to use a corrugated divider thatlocks the pot within the box. Soil found inside the box should also be contained. Preferably, put the pot inside a plastic bag then secure the top part of the bag surrounding the plant's stem. In case the plant doesn't have any primary stem, you may place a piece of paper in-between the leaves, as well as over the soil before securing it to the flower pot. Likewise, to prevent leakage and spills, ship any vases or plant containers without water plus cover soil in strong pots and vessels. Also, the flowers need to be kept hydrated during the entire delivery period to avoid any withering.

3.6 MARKETING AND SELLING FLOWERS

Statistics show that about 27% of all homes in the US purchase fresh flowers once or twice a year. This shows there's a wide market for floral businesses to consider, not just in the US but other countries also due to growing demand for flowers.

Figure 3.5: Marketing and selling of flowers is an important task to flower cultivation.

Source: https://focusedcollection.com/159245432/stock-photo-florist-handing-flowers-to-customers.html

Opening a floral storefront and waiting for buyers isn't sufficient enough for the advancement of your business. With fierce competition coming from other business players, you must market your flower products and services cleverly in order to stay ahead of the pack. There are various marketing strategies particularly designed for the floral business, which shall fix your concerns if the business isn't picking up fast enough. They include the following:

Embellish your shop pleasantly from outside: This business is all about optics. You want passersby to take a quick look at your flower shop from the exterior while moving around, and grab their attention going towards your shop. Decorations will give a good firsthand impression on them. Once amused, those window shoppers will likely come back to purchase flowers from your shop, if not straightway then in the near future.To create a pleasant impression of your store, make sure you pay close attention to your door signage and window displays. Take a serious look at your entry post and assess its design for attractive features. Your signage design plays a crucial role in building an attractive frontal appearance of your store. If the existing window display isn't eye-catching and entry signage also not up to the mark, then consider redesigning them.

Improve value for your existing clients: Don't worry too much about drawing numerous clients too quickly to your newly launched flower shop.

Instead, concentrate on the clients who visit your storefront regularly. Those who actually walk into your store are your most valued customers, so pay closer attention to them. Maintain honesty and sincerity towards them. Converse with them personally to see how you may improve their buying experience. For instance, send a written thankyou note to them for purchasing from your store, or alternatively thank them through phone. You may also provide some price cuts to your loyal clientele. These measures shall help your floral business grow steadily within your community.

Take advantage of the online space: In the digital era, spreading your flower business through a website is crucial. Among the easiest and most affordable solutions for taking your floral business name far, is by simply creating a beautiful floral site. Don't forget that individuals shall be placing orderson your flowers through the web also, which is necessary for the growth of your business.

When creating a flower business website, ensure it has pages informing people more about your shop, the flowers you vend, house delivery services, costs, discounts, privacy policy and any other information that would be useful to them.

Equally, your website design should stand out from those of competitors. Exceptionally designed web pages will create a great initial impression on your clients. Also, ensure that web pages don't take too long to load since this may put some clients off. Ideally, have the images optimized, with a combined code for central CSS or the JavaScript folder, then minify CSS, HTML and JavaScript (Henderson, 2008).

A reputable graphic design company can help you build attractive website designs for your business, and at reasonable rates. There are 100s of web designers with fresh design concepts that you can try out. Furthermore, the use of social media for updating followers about your friendly floral prices, thrilling offers or unique discounts is important. In case your shop is planning a flower event, you can publicize it on the social media platform. It's also an economical way of communicating a word about your business. Create your special Facebook page, Twitter, or any other appropriate social media platform for your flower business. Then, add content frequently on the pages to keep people updated. Preferably, you should put up 5-6 tweets each day to inform customers about new developments in your business (Vaughan, 2019).

3.7 TRANSPORTING FLOWERS TO BUYERS OR EXPORTERS

Figure 3.6: Transporting flower products from one point to another requires adequate care, and without it, things may become pretty messy.

Source: https://floribunda.nl/

Flowers are very delicate in nature, meaning crushing a flawless bouquet of flowers is probable with improper handling. Here are some ways to avoid bruising or breaking of flowers during shipping.

3.7.1 Use a Proper Vehicle

Ideally, carry them inside a mini or full-size van where most, if not all, the back-seats have been removed. The total amount of room you'll require essentially depends on amount of flowers you want to transport. Ensure that the van has sufficient air-conditioning, and put it in the backseat where the flowers will be kept; cool air from the fan shall help in maintaining their freshness. Air-conditioned vans are usually adequate for those who plan on driving their flowers for around 12 hours or a lesser period. If you got a longer trip that needs to be made and extra flowers to move, then consider leasing or borrowing a van with a freezer instead. It will keep your flowers at a moderate temperature of around 30 ^{0}F. (Henderson, 2008).

3.7.2 Apply Water-filled Buckets

Fill up several strong, plastic, 5-gallon containers with water. This water should comprise about 1/2 to 3/4 of each bucket. Put the cut flowers standing straight in the buckets. In case your journey shall take two hours or more, remove off the lowest 2-inches of the florets' stems before inserting them inside the water-filled buckets, this way they can stay hydrated during transport. Arrange the flowers within the buckets based on height and sort. Keep tall floras in one bucket, while shorter ones can be kept in another, including delicate ones as well. Ensure the buckets fairly full of floras to prevent them from being shoved around during the journey, but don't forcibly cram the blossoms together since this will make the petals to get bent, scratched or removed.

3.7.3 Ensure the Buckets are Stable

Add small, 10- or 20lbs sandbags onto the bases of the containers having tall cut floras to stop them from tilting over, just in case you need to break strong or turn quick while driving. Yet another way to avoid the buckets from falling is clustering them close together inside the rear zone of the van. Place sandbags around the border of the bunch of buckets to stop them from sliding or tipping over.

For shorter trips:

- Keep them away from the trunk: While you may assume the trunk is an ideal place to put flowers, the truth is lack of oxygen may impede your flowers' capacity to thrive, which means you won't enjoy their freshness once they arrive at the appropriate destination. Nevertheless, if you must utilize your trunk, it must only be performed so if the outdoor temperatures are fairly mild. In case the conditions are too hot or cold, then the flowers might wilt or freeze.
- Avoid the dashboard: Never leave flowers on the dashboard during a drive. Even though it's a flat surface, the dashboard is still in direct line with the sun and this isn't particularly ideal for newly cut flowers. You definitely don't want your nice bouquet to end up faded and slumped. Besides, it can obstruct your view while driving which is quite dangerous.

3.7.4 Packing Tips

Preferably, lay the groups of cut flowers neatly on their sides, in case you don't have immediate access to a van where you can fit the containers, or if you just have to travel for 1 hour or a lesser period. To move flowers in a car, stack them sideways on the rear seat. Place the batches atop each other, meanwhile taking heed not to destroy their petals. As for flowers with petals thatbreak easily, like lilies, these should be set on top of less-susceptible floras, such as Hawaiian ginger. Plus, you can apply an air conditioner for preserving coolness. For floral arrangements, place flowers thathave been packed in vases, baskets or any other container in much smaller, 3- or 5-gallon buckets (Henderson, 2008).

3.8 MAINTAINING AND REPAIRING MACHINERY SUCH AS PUMPS AND OTHER EQUIPMENT

Figure 3.7: Equipment like ventilation, lighting and watering systems found in a flower garden should be well-maintained to ensure quality yield.

Source: https://insteading.com/blog/greenhouse-materials/

Many machines and equipment are vital in achieving a successful indoor or outdoor flower farm. Just like the individual parts of an engine, these factors contribute to keeping the flower garden running smoothly. In case one component isn't working, it may compromise the whole outcome of your flower garden. That's why it's central for horticulturists and flower hobbyists to ensure every contributing element is operating properly. Therefore, it's important to have a maintenance checklist, such that nothing is omitted or overlooked. Some primary factors that should be looked at include:

3.8.1 Ventilation Systems

Ventilation units are commonly used in protected flower farming. The system is responsible for supplying fresh-air into the structure and removing putrid air out into the atmosphere, this component further controls temperature and moisture in the area. Equally, unless the indoor flower garden has a separate CO_2 injection system, it's the ventilation structure that also serves as the central supply of carbon dioxide for the plants (Vikaspedia, 2019).

However, since ventilation equipment serves a myriad of purposes in the garden, they are prone to damage from constant wear and tear. Maintenance of these appliances should be multifaceted so that every component is taken care of and nothing is left out. A daily checkup of the fans is essential to ensure that they're operating well. A broken fan will quickly cause a disastrous fluctuation in temperature levels which could be deadly to the plants.

It's good to touch the exterior of the ventilation motor to ensure there areno overheating or mechanical abnormalities. In some cases, a fan's motor may overheat before it entirely ceases working. A fan motor that's exceedingly hot to touch might be a sign of a failing fan. As an indoor grower, it's also a great idea to cleanse all ventilation fans following every garden cycle. Scrubbing the dirt, dust, or debris from the ventilation equipment will help improve the equipment's durability. Ducting should as well be inspected occasionally to ensure all connections and fixtures are snug and functioning properly.

Filters are yet another central factor for indoor floriculture. Both your intake and removal filters must be cleaned frequently or replaced to preserve the highest air flow possible. Unclean filters can cause devastation on fan motors since they generate increased resistance. While ventilation fans are constructed to withstand resistance, jammed down fans are highly likely to develop motor issues compared to fans running at lowered resistance levels (Vikaspedia, 2019).

For maintaining high- performance levels, air conditioners and filters must have their filters cleaned repeatedly. The air conditioner's coolant amounts should as well be inspected and maintained regularly, in order to ensure they're running at full capacity. Just like fans, jammed down air-cons are also likely to fail when poorly maintained.

3.8.2 Irrigation Components

Obstruction of emitters in flower irrigation systems should be minimized; this will save you from replacing clogged emitters or whole laterals which might be costly. Basic maintenance processes can be practiced to avoid jamming caused by algae, dirt bacteria, organic residues and clay particles.

3.8.3 Filter Cleaning

Pre-season, methodically clean all filters. Don't overlook the housing or body of your filters either. These housings may still get contaminated with sludge or scale when just the disc, filter mesh, or media is cleansed. For sand-filters, they must be back-flushed till completely clean, thereafter run off across the bypass until clear water is obtained. Cautiously inspect self-cleaning pressure screens for any technical faults, or holes inside the mesh, before putting them within the water supply. Additionally, once filters are washed and re-assembled, it's prudent to flush the structure by switching on the water then releasing the ends of sub-mains. These should be unbolted one at a go so that the fluid velocity rubs the pipe of sediment which has remained in the pipe. Shut off each submain once clean water begins to flow, then open the subsequent submain. Following the flushing all submains, begin to flush laterals copying a similar procedure. Several laterals can be released at a time, though ensure a sufficient skirt of water streams out the periphery of each lateral (Vikaspedia, 2019).

3.9 HIRING AND TRAINING WORKERS, AND ENSURING THAT THEIR WORK IS OF A HIGH STANDARD IN FLOWER FARMS

A floriculture worker is somebody who maintains and looks after flowers in the garden. They usually work under some form of supervision from a farm manager, where they might also receive onsite training. These individuals may provide service on flower farms of different sizes, varying from small, family-run companies to large-scale commercial operations.

3.9.1 Hiring Criteria

Since a good part of this job is manual, flower workers must display the ability to work outside in all types of weather. Additionally, those who operate as flower breeders may move from one farm to another, consulting with growers, and farm managers concerning their flower products (Larson, 2012).

Agricultural workers' must be ready to work long hours, harvesting blossoms by hand, while frequently crouching to cultivate and work the lands. They also lift up and carry harvested blossoms in baskets, and various agricultural tools in their daily course of duty.

Laborers also use pesticide sprays frequently on crops, where the levels of exposure and contamination are considerably high. Nevertheless, exposure can be reduced if appropriate safety measures are followed. The workers must be trained on how to handle chemicals used for controlling pests, without putting themselves at risk of contamination. Similarly, some flower farming machinery can result inserious injury, therefore workers should be continuously alert when using them to avoid any potential hazards caused by carelessness.

Typically, flower farm workers operate as casual laborers, where they move from one location to another as different flora ripen. Most of them have seasonal labor schedules, where they toil for longer hours in planting season or harvesting period when mature blossoms need to be picked from the farms (Larson, 2012).

3.9.2 Training and Welfare

Employees require constant training as planting technology continuously changes. When they are properly trained, it leads to reduced errors and work-related conflicts. Since this work can be tedious, employers must provide a favorable working environment for their workers and be aware of their welfare (Larson, 2012).

There should be a mutual understanding of the workforce for the smooth operation of floral businesses. Plus, they must be knowledgeable or cognizant of any existing labor laws, while following them accordingly.

The emphasis is on achieving a friendly working environment that is safe for everyone. Additionally, the practice of entire families, excluding underage children, working together on flower farms where they also live is becoming popular. It helps to promote the spirit of team effort, where workers understand that they need to cooperate in order to achieve something (Larson, 2012).

Nevertheless, flower farm owners should be careful not to encourage child labor, which is quite rampant in the agricultural sector. Childhood labor is reducing significantly on flower farms, thanks in part to activism from social welfare organizations. Bodies such as the International Labor Organization (ILO), working mutually with employer welfare associations, has numerous initiatives in the flower industry that have helped abolish the vice in this sector. State agencies also enforce the laws against this act quite strictly.

Many flower laborers are honest and work hard to make a living, still, in any society, you won't miss a few individuals who are dishonest. Farmworkers are also entitled to various benefits like maternity and paternity leave, transport allowance, and annual leave just to mention some.

Equally, some flower farms provide accommodation as briefly highlighted above. The employers may have collective bargaining agreements with their workers to provide these amenities and services that make life easier. Others even offer lunch. Generally, training of employees and providing them with social support services leads to higher productivity in the flower farm. Workers will be encouraged to deliver their best for the company since they are being treated well by the management (Larson, 2012).

Additionally, flower farm workers are encouraged to join a local welfare organization for employees. This way, they can champion for collective interests affecting them as a group such as pay increment, health care covers and right to legal representation in case of any misunderstanding with the employer. Membership to these organizations is usually open to everybody, so long as they are employed in the flower business.

3.10 ENSURING THAT HEALTH AND SAFETY, AND OTHER REGULATIONS, ARE COMPLIED WITH IN FLOWER FARMS

People employed in the flower production industry are susceptible to a broad range of hazards. Case in point, those who take care of and harvest crops in the open fields often suffer from heat-related disorders, pesticide poisoning, or chronic shoulder and back injuries coming from constant leaning, reaching, and lifting. Musculoskeletal conditions may also arise from intense repetitive work. Constant exposure to such factors can result in high rates of injury and fatality in the flower industry. Even so, there are various safety measures that flower farms can apply to ensure the safety of their workers, and avoid paying out hefty health insurance funds to employees injured in their normal line of duty. Below are some approaches that can be used:

3.10.1 Integrated Pest Management (IPM) for Controlling Chemicals Exposure

IPM is an approach whereby multiple modalities are applied in controlling pest invasions, like cultivation techniques, chemical plant protection agents, crop rotations and other similar practices.

By alternating pesticide use with other approaches, workers will limit their exposure to hazardous chemicals thatmight otherwise be lethal, especially to those who work long hours in the flower farms. Equally, populations in the surrounding area will be protected from spray drift or chemical spillage, which can affect the general environment where the flower farm is located. Substitution of hazardous pesticides with low-risk pesticides may further be pondered on a regular basis (Larson, 2012).

3.10.2 Providing Health Services to Workers

Medical tests on employees must be encouraged, plus those handling chemicals must undergo examinations at subsidized rates. Flower farm managers should be supported by relevant safety authorities to report incidents of accidents or occupational illnesses. This will essentially help in combating possible causes of worksite accidents, including provision of control measures (Singh, 2006).

Moreover, since most flower farm duties are done by women, usually they have greater incidences of injury and diseases compared to men plus are inadequately reached by healthcare services. Many of them have very little to no education, professional training or easy access to information concerning the risks pertaining to their work. Hence, the health services must be availed in special medical schemes involving the flower farm owners.

Statistics indicate that the rate of lethal work-related injuries among farmworkers is 7 times higher, compared to the ratio found among those working in industries like mining and construction. Additionally, the ratio of nonfatal job-related injuries is considerably higher among laborers in flower production jobs, mainly with respect to cases requiring several days off work or limited duty.

A sustainable work environment must incorporate fair and equitable job practices, alongside goals of safety, environmental protection, as well as welfare. Any health gaps among workers can be settled through policies that ensure labor laws are protracted to all laborers, notwithstanding their skill set or contribution to the business (Singh, 2006).

3.10.3 Training of Employees

Many flower workers are actually uninformed about their obligations and responsibilities at the farm. Hence, they fail to conform to safety legislations Occupational-Safety and Health act which are in existence. The predominance of seasonal or casual labor, alongside the extra limitations of illiteracy is commonly experienced in rural areas. Providing skills on occupational safety, including health is vital in ensuring that the laborers in flower fields are aware of the risks they are exposed to. Local Safety and Health officials may conduct training in all planting schemes within their area, providing workers with the information they need to stay protected in their day to day activities.

Floriculture managers should also be completely informed about the possible risks to laborers, including OSHA safety requirements that they must follow to ensure everyone is secure inthe field (Singh, 2006).

They must ensure young workers are given adequate training in safe work practices, plus demonstrate their capacity to perform duties safely prior to being assigned any task. These young flower farm employees should further be closely supervised so that any unsafe job practices can immediately be corrected. Also, it's important that workers are given protective clothing to prevent contamination by pesticides or other toxic compounds found in the field.

3.11 ENSURING MARKET CERTIFICATION SYSTEMS (SUCH AS QUARANTINE REQUIREMENTS) ARE UP TO DATE

Statistics show that agriculture constitutes around 38% of the globe's lands, out of which floriculture takes a considerable portion. Industrial agriculture isone of the potential environmental polluters needs adequate regulation, which can best be achieved through market certification systems. Flower farms are beneficial to the community, plus by minimizing their environmental footprint, you can safeguard adjacent lands and waterways, including flora and fauna. Not to mention they generate income for families, provide jobs to the local community, and play a crucial part in local conservation efforts.

3.11.1 Quarantine Policies

Quarantine regulations established by flower industry authorities are intended to control the introduction and transfer of contaminated blossoms, including other related planting materials and items, with an aim of preventing the inclusion of associated pests, weeds and pathogens to a region where there's no contamination. It's often applied in connection to suspected plant disease or illness, curbing the spread of these infections to the surrounding. Typically, quarantine is justified only in cases where the pest doesn't have a natural way of spreading, or when they're based upon biological considerations only, such as pest/pathogen inclusion risks and the existing safeguards (Ravinath, 2007).

Generally, risks are greater with the introduction of vegetative flower propagules as compared to true seeds. For the case of seeds, risks are experienced with deep-seated contaminations than with surface-borne infection of pests/pathogens. Also, risks are much greater with pathogens such as viruses, smuts, downy mildew and some forms of bacteria passed inside the seed, but without showing any external symptoms.

Once vegetative flower propagules are introduced, the rooted crops and other below-ground plant parts such as runners, rhizomes and suckers bear greater risks compared to unrooted cutting, budwood and scions. Whatever the case, making bulk introductions is always risky since thorough investigation and treatment for such cases is quite difficult. The planting zone may also be far too wide to prevent the establishmentand spread of introduced pests or infections (Ravinath, 2007). Based on these elements, plant quarantine provides the introductions as below:

- Total embargo or prohibition: In case the pest infection risk is high, or safeguards found in the country aren't adequate then import is prohibited.
- Post-entry quarantine:The risk factor is quite high, though adequate safeguards in form of post-entry seclusion growing facilities may be available.
- Restricted: Pest risk isn't high, plus import permit is needed specifying conditions for entry, assessment and treatment.
- Unrestricted: Here, an import permit isn't needed, and material could enter without restriction.

When formulating quarantine policies for floral products, local geographical conditions such as crop spectrum or environmental conditions might also be considered. Given that quarantine regulations are modeled to disrupt the life-cycle of pests or pathogens involved, any presence of secondary or collateral hosts within the nation of import including their introduction must also be put into consideration (Ravinath, 2007).

3.11.2 Quarantine Methods

Chemical treatments: Chemicals are either applied as spray, dust, slurry or dip. It's also important that the dosage of chemical is enough to eliminate the inoculum, though should not exterminate the host plant and the compound shouldn't be harmful to workers handling the treated seed either. Equally, the choice of which chemical to use and dosage should be made based on the particular pest/pathogen involved. The seeds treated at origin aren't just tough to examine but may be hazardous to inspect as well.

Tissue culture: Tissue culture, being one of the common safeguards in quarantine, reduces the pest/infection introduction risk in multiple ways. First, the consignment size is incredibly reduced as the introductions are characterized by meristem tips, expunged buds or embryos, plus secondly the aseptic plantlet structure has integrated pest/pathogen detection abilities. In tissue culture, all bugs, mites, nematodes, including most fungi can easily be eliminated. Any symptoms on developing seedlings, or growth of the microbes on the agar part, if any, might be visible via the see-through culture tubes, and they could be discarded. Generally, when tissue culture is combined with thermotherapy or chemotherapy, it makes an excellent quarantine safeguard for flowers (Ravinath, 2007).

3.12 KEEPING PRODUCTION AND FINANCIAL RECORDS AND WRITE REPORTS

Every successful flower grower can prove the significance of keeping track of their business's expenses and revenues. As people continue warming up to the prospects of becoming floricultural entrepreneurs, there's a crucial need to teach them on primary bookkeeping and accounts. Notably, every farmer, despite his farm's scope or activities therein must have a broad insight intothe economic status of their business.

There's no better way ofgaining such a perspective than keeping production and fiscal records. Bookkeeping further acts as the basis whereby the grower can provide future management decisions concerning the farm, or seek for loans from financial bodies. Typically, financial institutions only provide their credit to flower farms that display an ability to repay. That capacity is best demonstrated through suitable bookkeeping practices (Ravinath, 2007).

It's vital for every flower farmer who's serious concerning their agricultural business to have good and precise records. These records shall be helpful in coming up with the correct management, marketing and improvement decisions. From the records, a grower can determine their efficiencies and inefficiencies, plus based on similar parameters, they can calculate the farm's gradual progress or its lack thereof. Furthermore, the financial records give a farmer some idea ofwhat improvements they need to make.

Preferably, good farm accounts should indicate when a flower was planted, resources used in land preparation, planting, reaping, and marketing, including the returns gotten after selling the flower produce. By subtracting all costs from the sales amount, a grower is able to determine his return or loss margin. Particularly, bookkeeping and accounting don't need to be complicated. Those who have computers may use simple accounting packages like Excel, whereas others may favor the written method of bookkeeping. It's just an issue of personal preferences though. Whatever approach one prefers, nonetheless, it's crucial to document whatever amount you spend or gain from your flower business (Ravinath, 2007).

3.12.1 Treat your Land as an Asset

Properly managed floricultural land shouldn't depreciate in value but rather go up. However, mismanaged land may take several years of precise nurturing to go back to productivity, particularly if it's getting highly acidic or there's drainage of nutrients caused by over-farming in the past.

The cost of preserving your land must readily be accounted for. Soil pH management is crucial. Different crops require different pH amounts. Acidity usually means more weeds, even though some plants may need more acidic soils. A keen analysis is required for adding the correct nutrients and maintaining soil pH at proper levels. Also, weed removal, regardless of its manual removal or spraying is an ongoing cost (Ravinath, 2007).

If you're growing flowers, you'll also need to control pests. It's a continuous battle with them however since some pests are known to develop resistance over time when exposed to certain pesticides. The same applies to fertilizer since it's an inevitable cost if you must keep your flower fields productive. If looked up well, top quality land should stay productive year on year. Therefore, whatever it costs for keeping your floricultural land in good shape it's likely that money will be spent. Ensure that you keep financial records covering all these expenses.

3.12.2 Stay Updated with State Subsidy Programs

Floriculture is a critical field as it contributes to economic growth. For that reason, a number of governments provide subsidy programs to farmers to aid them out in their tough years, or to make flower farming more appealing for those who would want to try it out. These subsidies are diverse in each nation and they usually change. In one year, you may find there's a huge subsidy on greenhouse flower production, yet another year it could be field growing that's subsidized (Ravinath, 2007). Sometimes though governments get things wrong, causing surpluses thatdrive down floral products costs too far. Today subsidies are handled more efficiently, though that could mean they alter more often too. Therefore, ensure that you keep record of all subsidies and properly account for them, particularly if they're done as direct payments. The greater the knowledge you have on subsidies, the better you can plan for your flower farming strategy and make the good of it.

3.13 RESEARCHING GROWING AND HARVESTING TECHNIQUES, AND NEW VARIETIES

Scientists all over the world are seeking to improve flower quality and production by engaging in continuous research. Most agree that florets need adequate sunlight and well-drained soils in order to grow their best. Research shows that they generally do great in low to medium-fertile soils, therefore avoid amending the soil too much with organic material before planting. However, you can still add lime in order to increase soil pH to about 7.0, which is the most ideal level for growing blossoms.

3.13.1 Growing tips Throughout the Season

Most flowering plants like perennials grow to an average height of 1 to 3 feet tall, plus some don't require staking. Watering is crucial for young plants and needs to be done on a daily basis, preferably with the help of sprinklers. Upon establishment, some flowers are drought-resistant and don't require regular watering. But growers need to be careful of over-watering as well, as it's a common source of stress in flowering plants (Singh, 2006).

Similarly, in matters fertilizer, apply it moderately so that weeds don't get the opportunity tofeed off the nutrients as well. For some drought-tolerant flowers, they wouldn't need mulching to preserve soil moisture, apart from in extreme cases. If you decide to mulch, then use a small-sized bark plus don't allow it to pile up near the crop's crown or your flower might rot.

However, in northern territories, mulch your flowers during late fall in order to shelter them from winter's freeze. Stack a 1-feet deep mound of bark mulch or wood chips on the crops after a chill. This shall protect them from cold, though not make them rot. Take out the mulch during early spring.

For trimming and pruning, research shows that most flowers blossom in summer. During this period, flower stalks may be harvested and utilized fresh or dried. If you aren't gathering lavender floras to use, still cut off spent blossoms soon after the floras fade in order to embellish the plant and promote a second flowering (Singh, 2006).

When it comes to woody ornamental plants, they produce their finest and most aromatic foliage and floras from young stems. Ideally, prune 2-year or older plants during the spring season, chopping the wooden stems back by around one-third. This shall encourage new growth, which results in improved foliage and flowering.

3.13.2 Harvesting Methods

Most scientists agree that harvesting of cut flowers should be done during the early morning hours after the dew has dried up. Respiration, a factor that influences the quantity of water a flower should be given, is usually low during the early morning. Respiration rate doesn't just influence cellular turgidity; it also affects carbohydrate intake. Thus, harvesting flowers near to dawn when the respiration is minimal will ensure that they last for longer.

Immediately after plucking your flowers, put them in water immediately to provide humidity. Have a pail of water around when you harvest the flowers. Because cutting flowers split them away from their main source of food and water, having a water bucket handy and ready for use is essential for maintaining freshness until the cut flowers are ripe for transfer into vases. If possible, flowers must also be put in a shaded area to help with controlling water loss.

3.13.3 Flower Details

It's important for growers to know specific details about each species, including the variety of flowers they have. Researchers recommend that all flowers must be harvested during the desired maturity period, which varies from one breed to another. Blooms are 50%, 75% or 100% open. Some flowers are well off if harvested while blooms are around 75% open, whereas others will persist to open blooms in case harvested at 50% open (Singh, 2006).

3.13.4 Pest and Disease Management

Since flowers are very fragrant, pests, including insects, powdery mildew or fungus might be a problem. However, you can prevent them from causing havoc by applying safe pesticides and spacing crops further to prevent moisture accumulation, which is a breeding condition for fungus. It will also promote good air circulation for keeping the leaves dry and rot free (Singh, 2006).

4

PROPAGATION METHODS OF FLOWER CROPS

CONTENTS

4.1 Introduction 90

4.2 Raising Flower Crop From Seeds 91

4.3 Raising Flower Crop Through Vegetative Propagation 103

4.4 Raising Flower Crop Through Tissue Culture 109

4.1 INTRODUCTION

Flower propagation refers to the process of reproducing or increasing the number of plants of a given species or cultivar. Generally, there are two basic forms of crop propagation, which include sexual and asexual. Naturally, propagation of floras most often occurs through sexual reproduction or creation of viable seeds. In exposure to suitable environmental conditions, the seeds sprout and grow into complete, reproductive plants. Throughout the years, florists have also established asexual propagation techniques that use vegetative crop parts. This permits plants to be formed in ways which nature can't duplicate.

Figure 4.1: Propagation is a biological process used by growers to reproduce and multiply their flowers.

Source: https://coffeetocompost.com/2014/07/21/propagating-torenia-wish-bone-flower-from-cuttings-more-free-plants/

4.1.1 Sexual Propagation

This propagation technique uses seeds to generate new crops. Sexual propagation is fairly easy and often needs no special tools or equipment. It's nature's means of obtaining several plants economically. Viable seeds are developed once pollen is moved from the flower anther to the stigma where fertilization occurs. For self-pollinated plants, every seed created will have the same genetic code as the parent plant, preventing mutations.

4.1.2 Asexual Propagation

Asexual flower crop propagation applies vegetative sections of the flower to reproduce a clone, which is a precise genetic replica, of the parent plant. It can have various advantages, such as the ability to multiply genetically the parent crop. Inmost cases, it permits for floras to be matured to a size that's appropriate for transplanting within a shorter period than from seed. Similarly, certain plants create seeds thatare sterilized or have low viability, making sexual propagation hard or impossible. Growers use several techniques to propagate crops asexually by;cuttings, division, layering, grafting or budding and creating new crops from tissue cultures within the lab (Milio, DeMuro, and Martin, 2017).

4.2 RAISING FLOWER CROP FROM SEEDS

4.2.1 Germination of Flower Seeds

Flower seed germination occurs when a healthy seed, with adequate internal properties, is exposed to positive environmental conditions. Some internal factors that impact on seed germination are seed coat elements and dormancy. During the process, reserve starch found in the seed is hydrolyzed to deliver fuel for energy, including building blocks for fresh macromolecules. Various environmental factors influence seed propagation such as humidity, temperature, oxygen and, lighting. Appropriate germination conditions vary depending on the species, therefore it's crucial to know your plant well. Usually, information regarding the germination period and other crop characteristics arewritten on seed packets (Milio, DeMuro, and Martin, 2017).

For faster maturity and improved performance, crops of various annual flowers and ornamental plants are usually started indoors, just before the outdoor growing season. The appropriate sowing period before outdoor transplanting time varies from 4 to 10 weeks, subject to the flower species. Successfully growing seedlings indoors needs a few distinct considerations.

Seedling care: After the seeds sprout, the seedlings should be given appropriate growing conditions in order to ensure sturdy, vigorous crops for outdoor planting. Put them in a bright place, preferably south-facing the window. Similarly, a fluorescent light fitting with a pair of 40-watt bulbs may be used; place the light unit about 6 inches over the seedlings then switch it on for around 14 to 16 hrs daily. Annuals and vegetables typically prefer night temperatures ranging between 60 to 65 ^{0}F whereas daytime conditions maybe 10 to15 degrees higher. Additionally, always keep humidity levels around the seedlings relatively high. You can begin fertilizing them using a liquid-soluble fertilizer, measuring about half the recommended dosage just a few days following the seedling's budding, then regularly thereafter.

Fertilizing application rate depends on factors such as species, fertilizer concentration and environmental factors. Always maintain strict hygiene standards when handling flower seeds, this is central for illness prevention caused by insects or microbes that may be hiding within the seeds. Among the most common infections is "Damping -off," though it's easy to manage through drenching the seedlings using a fungicide, applied at half the required rate for mature crops (Milio, DeMuro, and Martin, 2017).

Soil basins: The basins for developing seeds must be clean and strong, also fitting into space at hand for germinating plants indoors. A proper container aids in giving the flower seedlings a good start, plus it saves on overall work. Items like flats, plastic trays, and pots may be used for this purpose plus they work quite well. While clay pots can be handy, they aren't recommended since they're rather hard to disinfect between usages. Contrarily, peat pots and compacted peat pellets are common bio-degradable choices that permit the container to be grown alongside the seedling, thereby avoiding the shock of transplanting. Also of importance are seed coats, some species of flowers produce seedsthat are almost impervious to water. Naturally, this feature tends to provide longevity to seeds, though it may further make them less probable to germinate soon after being produced. Gardeners may further encourage seeds with solid seed coats to sprout by adjusting their protective casing.

4.2.2 Factors that Control the Germination

Age: As flowering seeds age, they tend to consume nutrients kept inside them at a fast rate. Once all the nutrients have been expended, the embryo will then have to germinate or wither. Some seeds lose functionality rather quickly, while others have crops with well-developed survival systems. Nevertheless, most desirable flower species tend to have reduced seed lives.

Storage: Seeds shall stand a greater chance of growing if they've been stored properly. All seeds sprout more effectively when kept in cooler and dry conditions, therefore you should always be wary of those seeds which are stored in contrary conditions, warm and moist. During seed examination, discard the whole packet if it shows aspects of mold or other related fungus (Milio, etal, 2017).

Most commercial seed packets also have a "use by" period printed on them. However, don't take the date too seriously—since seed manufacturers apply this date to provide customers a large proportion of germination, though many seeds shall remain viable for several years following the date written on the packet. Nonetheless, the printed date would give you some sense concerning how old the seedling packet is. If you're just a year or so beyond this date, then there's a good probability majority of the seeds shall still grow when planted. But in case the seed pack has stayed for 6 years or more, you should expect to get a much lower ratio of germination. Proper storage practice is to date your seed packet upon buying it, to ensure you'll know precisely how old the seeds are when you want to use them in the future.

Stage of maturity and viability of seeds: The maturity of seeds is typically determined by key interactions, occurring between diverse hormone signaling paths, including other regulatory cues availed by the seed's environment. Clarification of various important events that control seed maturation is definitely among the most exciting aspects of seed biology. Collectively, they are crucial in determining various processes that ultimately control seedling vigor, such as protein and oil deposit accumulation, management of precocious germination, dormancy induction, and creation of stress tolerance. Embryos extracted from growing seeds at certain phases often germinate, when put on water or any minimal culture medium. Therefore, while they haven't yet attained their full development, embryos grow precociously, though often fail to form viable seedlings.

Thus, the vital role of any seed environment is maintaining embryos during their developmental stage, until they're wholly formed and have gathered sufficient reserves to allow successful germination and successive seedling establishment (Singh, 2006).

Research to understand the mechanisms in which the seed environment sustains the embryo, especially during the developmental stage, while suppressing growth have implicated various regulatory factors, such as abscisic acid and inhibited water uptake. One element of the chemical ecosystem surrounding the growth of seed embryo is osmotic conditions. A suitable floral growing ecosystem should allow for high negative osmotic factors during development, consequently this is believed to constrain precocious growth of embryos by controlling water uptake. Basically, a large part of development or growth of the seedling embryos, including seed storage tissues relies on the establishment of reserve materials. The seedlings capacity to emerge and get established is also significant.

Water: Flowering plant seeds are usually encased inside the outer shell of different thicknesses. This shell safeguards the small embryo found inside from parasites, damage or extreme temperatures. Within the shell also lies the endosperm, which nourishes the developing embryo with food during its initial growth stages. The shell absorbs water, stimulating the biochemical processes needed for growth and germination. Upon absorbing sufficient water, the embryo breaks the seed's outer casing and the tiny, growing crop emerges.

The hard shell present in some seeds is made to inhibit water from going to the seed, thus stopping the embryo from accessing the water needed to activate the growing mechanism. Hard flower seed coats soften once soaked in water, including when the soil gets consistently moist, though unless there's some method to transfer water to the seedlings, they might die before even germinating (Singh, 2006).

Thin seedling coats remove the water barrier. Consequently, light passes through the coverings of thin crop tissue, plus when the pigmentation phytochrome absorbs light, then the embryo is changed in a way whereby it appears from the thin seedling coating. Slightly-coated seeds require water so they can intake nutrients from the soil. Scarification is the procedure of nicking tough-coated flower seeds to extract a tiny portion of the external shell. This permits water to flow into the area and access the embryo. The greater amounts of water the embryo intakes, the larger it swells, eventually bursting the whole hard external shell after germination. Nicking further makes it conceivable to see if a flowering seed is viable since healthy seeds look white below their coating.

Supply of oxygen: Usually, seeds stay dormant or inactive within the soil until appropriate conditions for germination present. Among these is oxygen, which is necessary for catalyzing food intake and providing ample conditions to foster germination. Oxygen is essential for respiration that releases the energy required for growth. Growing seeds respire quite actively and require sufficient oxygen. The growing seeds get this oxygen directly from the air found in the soil. It's for this particular reason that the majority of seeds sown down in the soil or water-logged conditions where oxygen is deficient fails to sprout due to inadequate oxygen. Tilling helps to aerate the soil thereby facilitating good germination (Singh, 2006).

Once a seed is exposed to sufficient oxygen, the embryo's cells would start to enlarge. Thereafter, the seed coat cracks open and then a radicle or root emerges first, closely followed by the plumule or shoot bearing the leaves or stem. Various factors can cause poor growth. Overwatering makes the crop not to get enough oxygen. Equally, planting seeds too down causes them to utilize all of their kept energy before breaching the soil surface. Dried conditions mean that the crop doesn't have adequate moisture for the germination to keep going, despite oxygen levels being sufficient. Furthermore, some flower seed coats may be so tough that oxygen can't get through unless the coat is broken down. Soaking or rubbing the seeds shall help soften the seed coat so that more air can pass through it.

Nevertheless, for storage purposes tightly sealed containers are recommended. Studies show that air containing 21% oxygen or more may reduce seed quality or germination capacity during storage.

Heat or temperature: Seed germination is a delicate event thatinvolves various factors like temperature. Propagation increases in elevated temperatures, but up to a given point. After the seeds reach optimal temperatures, depending on the flower breed, then germination starts to drop. Some crop seeds, such as cool-season ornamental plants and flower rhizomes, will sprout best in temperatures of between 55 to 70F. Whereas warm-season plants will germinate best at temperatures of around 70 to 85 F (earth care greenhouses, 2015).

Therefore, regardless of whether it's high heat or cold, the temperature will affect flower plants including their growth patterns. It's among the reasons it's so crucial to check a crop's hardiness and decide whether the plant is compatible with your specific growing zone. High temperatures will affect seed growth in multiple ways. The obvious one is on photosynthesis, whereby floras use carbon dioxide to generate oxygen, and respiration, which is the opposite process where crops use oxygen to create carbon dioxide. Scientists explain that both procedures increase when temperatures increase. Nevertheless, when temperatures go to uncomfortably high limits depending on the plant, these two processes may become unbalanced.

Some flower seeds, for instance, get into problems when temperatures surpass the 96 degrees F limit. Effects of temperature on crops vary widely, mostly influenced by factors like sunlight exposure, including differences between night and day temperatures, plus proximity to adjacent rock structures where thermal heat mass may originate from. For outdoor seed growing, consider covering the mulch using a coating of frost safeguard fabric or some old blanket, then hanker it down to avoid blowing the covering away (earthcaregreenhouses, 2015).

4.2.3 Factors Limiting Germination of Flower Seeds

Seed dormancy: Seed dormancy is described as a state whereby seeds are barred from sprouting even under environmental situations normally favorable for growth. These factors are a complex blend of light, temperature, water, gasses, seed coats, hormone structures and mechanical restrictions. In the flower sprouting industry, dormancy is usually considered as a seed generating plant's capacity to endure cold weather. The more "winter hardy" or dormant a seedling is, the better placed it is to grow and survive during the cold winter season (earthcaregreenhouses, 2015).

Some of the different dormancy ratings are, very dormant, dormant, and moderately dormant. As a practical rule for sprout producers, highly dormant seeds are typically easier to cultivate without having spoilage issues, particularly in the summer. Nevertheless, they typically have tougher seed, which might take much longer to sprout, plus a lowered yield for some flower crops. For commercial growers to make best of their profits, they must balance the rate or pricing of the seed together with the form of seed they require and expected yield.

Most flowering seeds will sprout and grow without any encouragement, other than interaction with humid soil, whereas others, mostly woody ornamental plants and wildflowers, shall remain dormant up to the right time. Naturally, environmental triggers alert dormant seeds when it's rife for them to come up.Within a gardener's care, flowering seeds are typically stored under controlled settings so that they don't encounter the effect of changing seasons, temperatures, light and other features thatshow them it's the best time to germinate, therefore they remain dormant (earthcaregreenhouses, 2015). The seeds then require special treatment to propel them into believing their magical moment has arrived.

Stratification: During stratification, seeds are exposed to cold temperatures inside the refrigerator, triggering a short winter condition. Once brought back to normal room temperature, they sprout readily. Larger seeds are easier to stratify through placing them between humid paper towels, or inside a jar of humid sand within the refrigerator. In general, it's appropriate to plant small seeds inside a soil tray before refrigerating them. This process usually takes between 2 and 6 weeks of freezing temperatures to disintegrate the inhibitors, based on the specific type of seed.

Figure 4.2: Stratification involves putting seeds in frozen conditions so that they can germinate quickly once introduced to warm temperatures again.

Source: http://www.usa-gardening.com/cold-stratification/cold-stratification-3.html

- Lighting: Exposure to lighting disintegrates the germination inhibitors for some breeds of seeds, especially wildflowers thatproduce small seeds. For the best outcome, press seeds thatrequire light exposure on the surface of some humid potting device or garden soil, though avoid covering them with soil. The seeds might take a couple of days to sprout, meaning you shouldn't lose hope on them very soon.

Seed rest period: Some species of flowering plants have seeds thatdelay germination for several months or years, plus some seeds are known to remain inside the controlled seed bank for up to 10yrs or more, but still germinate when conditions are right. Others have a considerably long viability duration, where the oldest documented sprouting seed was almost 2000years old depending on radiocarbon dating.

- Physiologicalrest:Physiologicalrestingperiodpreventsembryonic growth and seed propagation until chemical fluctuations occur. Physiological rest is shown when a surge in growth rate happens after the appliance of gibberellic acid or GA3, including after Dry post-ripening or dry storage. It's also pronounced when latent seed embryos are expunged and develop healthy seedlings: or in situations where cold weather is experienced for up to 3 months at degrees of 0–10 °C, or warmer climates 15 °C that allows foster stratification: or in case dry after-ripening reduces the cold stratification duration required. For some seeds physiological latency is witnessed when scarification improves germination. The physiological rest period is broken when impeding chemicals are disintegrated or are not produced by the flower seed; typically, by a duration of cool humid conditions, usually below 39F, or for the case of several ornamental plant species found in the Ranunculaceae family. Abscisic acid is commonly the growth disruptor in flowering seeds plus its production may be affected by lighting. Some plants such as Peony species contain several forms of physiological latency or rest, one affects the root growth whereas another controls plumule or shoot growth.Equally, some ornamental plants including a couple of grasses, including those from perennially arid regions require a duration of drying before final germination, while the seeds are set free they still need a lowered moisture content prior to germination. In case the seeds stay moist after dispersal, then germination may be postponed for several months. Many herbaceous ornamental plants grown in temperate climatic zones have a physiological rest period, which typically disappears when the seeds start to dry.

- Morphological rest: In morphological rest, the embryo is not well-formed or undifferentiated. A few seeds have wholly differentiated embryos which need to mature more prior to seed germination, or sometimes the embryos may not be differentiated into varying tissues at the period of fruit ripening. For immature embryos, some flowering plants discharge their seeds right before the embryonic tissues have fully differentiated, plus the seeds ripen right after water intake on the ground, while germination can be deferred from afew weeks to months.
- Combined rest: For these types of flowering seeds, they have a morphological and physiological resting period at the same time. Also known as morphophysiological latency, it occurs when seeds having underdeveloped embryos, further have physiological factors to dormancy. These seedlings, therefore, need dormancy-breaking treatments, including some period of time for development of fully grown embryos. Closely related

to it is combinational rest that happens in some seeds, whereby dormancy is triggered by both endogenous and exogenous conditions. Some ornamental plant species contain both hard impenetrable seeds coats, including physiological dormancy. However, not all seeds pass through a period of rest, some species of flowering plants remove their seeds late during the year where soil temperatures are too low for growth, or in case the environment is arid.

Presence of hard seed coat: Seed hardening affects many flowering plants and is a common factor in germination. Hard seed is witnessed when the plant's seed coat becomes impermeable to water. In fact, hard seed typically is also a form of dormancy, since it prevents the growth of viable seeds since seed coverings are impermeable to moisture. The ratio of hard seeds is described as a fraction of the overall percentage germination in the flowering plants. However, for different crops you may see a completely separate category for the hard seed measurement (Henderson, 2008).

There are several factors that cause seed hardness:

- Sclerenchyma: Ornamental plants are quite different from other organisms in their structure or makeup. While animals rely on their skeletal system for support or flexibility, including performing normal life functions. Plants, contrarily, rely on their cells for specialization and performing their regular functions. Case in point, since plants don't have bones that allow for support, they need cell walls surrounding their cells to keep them upright. Sclerenchyma cells are unique plant cells thatexist to offer strength and support. They're found in all types of plants such as grasses, flowering plants and trees. Despite their usefulness, these cells can sometimes develop exceedingly tough layers that make it impossible for flower seedlings to germinate. These are typically strongest form of cells, surrounded by a secondary cell wall thatis toughened with lignin. This makes them both tough and durable. These cells are commonly found in crop parts thataren't lengthening any longer. Once the cells expire, their cytoplasm and nucleus degenerate through their cell walls stay intact, providing structural support for the plant. Similarly, sclereids are a denser form of the sclerenchyma cells, having highly thickened, lignified cell walls thatform tiny bundles of robust layers of tissue present in most flowering crops. The presence of multiple sclereids can produce a gritty texture on the surface of seeds, which further impedes their germination.While sclereids are flexible in shape, these cells are mostly isodiametric, forked, or intricately branched. They may be clustered into bundles, form holistic tubes found at the margin or can emerge as single cells or tiny groups of cells present in parenchyma tissues. Isolated sclereid cells are commonly called idioblast. These sclereids are usually found in ground tissue, vascular tissue and the epidermis of plants and seeds.

- Stone cells: Stone cells refer to a form of sclereid cells comprising of a dense secondary wall, also with an uneven shape. These vary in scope and thickness levels, plus are well-known for their branching format including irregularity. Sometimes, vascular bundles develop in the merging point of the center and inner mesocarp, consisting of scalariform, phloem material and spiral, including scalariform-reticulate vessels. You can easily identify stone cells from their awkward-shaped cells, which are spread throughout the seed's mesocarp. These cells typically vary in scope up to around 220 μm, they have a long axis and comprise of a yellow to brownish-yellow color. Stone cells may occur individually in the form of idioblast or within small groups. Nevertheless, when flower seeds are properly selected before planting they will likely not have stone cells.

4.2.4 Methods used to Ease Germinations in Flower Seeds with Hard Coat

Cracking of the coats by mechanical means: Generally, if your flowering seed doesn't sprout within the first 5-7 days then it won't likely do so at all unless a physical intervention like cracking is used. What will happen is that it will just ferment and become black, which is a sign of rotting. Nongerminating flowering seeds occur from one time to another, but there's no need to worry as cracking the coat can still make them grow. The effects of mechanical breaking depend on humidity levels, the moisture needs to be balanced so that the seed doesn't get damaged when being hit. When the seeds have less moisture content they usually are more susceptible to breaking. Likewise, external cracks are much greater in bigger seeds compared to their smaller counterparts. When performing this procedure, be careful not to inflict mechanical abuse on the product by applying too much force, this may cause internal injuries that may affect the crop's ability to grow naturally. In fact, this can lower the chances of your flower sprouting by around 60–70% (Milio, DeMuro, and Martin, 2017).

The following is the procedure:

- The first step is wrapping some paper towels around the seeds for protection. This material should be lightly moist, then folded multiple times until it fits the content inside.
- Upon folding, put your seeds in the middle of the towel and put in the bag, squeeze all the air out and seal.
- Next, put the seeds in a warm environment for a few minutes to make the outer shell easier to crack.
- Once splintered, with a tiny taproot showing, use tweezers to gently pick up the flowering seed, while paying specific attention tothe tap so as not to damage it.

Abrasion: Abrasion in floriculture involves weakening, scratching, or otherwise modifying the seed coat to stimulate germination. The reproductive tissues of many flower species are typically impermeable to water or gasses, therefore inhibiting or delaying germination. Abrasion or scaring is an effective way of solving this problem. This procedure is typically done mechanically, chemically or thermally. The seed's hard outer cover makes it resistant to air and moisture which aid in germination. To prevent this, you may need to slightly scratch, or nick the coat. Naturally, this occurs when flowering seeds move through the digestive system of animals, which consume them, including through freezing winter temperatures or bacterial activities that disintegrate the seed coat. Seeds thatneed to be scratched before sowing are generally large or contain thick seed coatings, similar to sunflower seeds.

The procedure of seed scarification may be accomplished through rubbing the seedling on file or sandpaper. Remain careful not to scratch too deep into the flowering seed itself. Do it just enough to see a color difference in the seeds, but without compromising its form or quality. Upon scarring, water is effortlessly soaked up via the seed cover upon doing this (Milio, DeMuro, and Martin, 2017).

There are numerous methods of abrasion such as soaking the seedlings in hot water, applying sandpaper or file and acidification, whereby a concentrated solution is used to scratch open the coating. This is a delicate procedure and you must keenly follow the necessary directions to prevent your seed from getting damaged. The kind of abrasion will largely depend on the kind of flowering seeds you're germinating.

Soaking in water or acid: Soaking seeds in water before growing is a popular gardener's trick which many flower farmers are increasingly adopting. When you put flowering seeds in water before growing, you can considerably reduce the duration of time it would take for the seedling to germinate.

Figure 4.3: Soaking is an effective way of softening hard flower seeds in preparation for planting.

Source: https://seedtocrop.net/2018/11/why-you-should-soak-the-seeds-before-planting-them-and-how-to-do-it/

Soaking is especially useful for flower gardeners with dense clay or sandy soils. Clay is hard to humidify evenly if it's let to dry out; plus, once it becomes wet, it can sustain humidity for too long, making seeds to rot. Conversely, sand drain quite rapidly, making it challenging to keep the seeds humid enough for sprouting to occur.In case the seeds are already humid, or the seed coats absorbed sufficient water for roots to emerge, it could mean earlier harvesting than when sowing flowering seeds in their dried state. It also eases some of the distress of guaranteeing that your soil is sufficiently damp after sowing a line of seeds.

In the wild environment, a seed usually expects to encounter severe heat and cold, or intensely wet and dry conditions, or surviving the acid-filled gastrointestinal system of an animal. But for the modern-day flower garden, seeds are relatively pampered. Soaking them before planting can help to disintegrate the seed's natural defenses, allowing it to grow much faster. Additionally, for many seeds, humidity levels have a significant role in determining whether the seed is optimal for sprouting. By soaking up the seeds, you can rapidly boost the moistness level around the seeds, and this signals the seed to grow safely (Singh, 2006).

Stratification: Seed stratification is the procedure whereby winter conditions are simulated in a seed, such as cold and moist, in order to prepare the seed for its usual germination temperatures which normally come during spring. It might involve placing the seeds in a refrigerator, typically in some form of humid potting soil. Alternatively, it could mean keeping seeds outside in winter inside an airtight plastic bag or enclosed container, often with some grow mix. Mostly, seeds thatbenefit from stratification are the small perennial flowers. Like delphinium and violets, which even when germinated indoors, have greater germination rates following stratification. In fact, many perennial flowering herb seeds are known to perform better following the cold, stratification procedure. In order for seed stratification to be effective, it's vital to mimic the specific conditions they require when overcoming dormancy in nature.

While some flower seeds do best in warm and moist conditions, others need cool and wet factors. More still, other seeds need a blend of both temperate and cool factors followed by some warm treatment, or a mixture of tepid and cool conditions, accompanied by a drier cycle and warm conditions to germinate. Thus, knowing which seeds need to break their dormancy is crucial before starting any seed stratification work. Sometimes, cold treatment for flowering seeds may be necessary for those plants that require more time in the soil over winter so as to germinate. Consider the minute ornamental poppy seed, for instance, which benefits from a slight chill period before growing. They should be refrigerated for a couple of weeks prior to planting in your garden (Singh, 2006).

4.3 RAISING FLOWER CROP THROUGH VEGETATIVE PROPAGATION

4.3.1 Cutting

Cutting propagation is an affordable and simple way to develop new crops from flowers that you already have. It's an asexual form of reproduction that produces plants which are genetically similar to their parent. Cuttings rely on the most spectacular aspect ofplants, which is their cell's capacity to duplicate various parts and aspects of the crop. By taking a leaf cutting or stem then creating the appropriate conditions, you may develop a completely new plant. Preferably, begin with a 'parent' plant or stock that's in good health and has considerable amounts of stems, such that if a stem is removed, it still won't harm the flowering plant (USDA, 2019).

Stem cuttings: Propagation through stem cuttings is usually the most popular crop propagation technique for ornamental plants and woody shrubs. It's also an effective technique for houseplants. The latter are generally quite easy to broadcast. Simply look for a fit stem without flower buds, infection or pests. Using a sharp, clean knife incise a clean cut at an angle of 45 degrees, in order to maximize the plant's rooting area. Cuttings must be around 3 to 6 inches long or include the stem's tip, plus roughly around 2 or 3 sets of leaves fitted.

- Take out the bottom section of leaves and put the end you've just sliced into a rooting gel. This shall help seal up the cut crop tissue and support fresh root growth.
- Thereafter place the cutting inside a small pot containing moist vermiculite, soilless potting mix or perlite.
- Meanwhile, poke a tiny hole in the germinating medium before inserting the cutting inside it. Through this method, the rooting solution won't easily get rubbed off from the plant's stem.

Ensure that your new floras stay warm and remain in bright light, though out of immediate sunlight. Most cuttings shall also benefit from extra humidity. To improve moisture levels and develop a mini-greenhouse condition, put the pots inside a transparent plastic bag. Don't allow the plastic which you apply for covering the pots to touch the cuttings. Items like milk cartons, mayonnaise jars, and synthetic soda bottles may also be applied to cover the cuttings. Upon developing roots, a process thatmay take some few days or months- take time to replant them inside another container with humid, though not wet, potting soil (USDA, 2019).

To determine if roots have developed or not, push lightly on the crops. In case they jet right out, then they still aren't ready. If you experience some resistance, just go ahead and re-pot once more. Unless the new crops have become wholly established, keenly monitor the volume of moistness and light they receive. Take out dropped leaves or diseased crops from the zone immediately they are detected, in order to prevent the fungus from diffusing into healthy plants. As for softwood stem cuttings, these are usually derived from new shrub branches thathaven't yet turned woody. The phrase "softwood" actually describes the phase of growth on deciduous woody crops that isn't fresh or green, but also not fully mature or woody. It's somewhere in-between these two forms.

4.3.2 Layering

Layering is a technique whereby a plant's branch having one node or more is bent onto the ground, whereas a portion of it is concealed with humid soil thereby leaving the apex of the branch visible above the ground. Within a short time, fresh roots develop will form from the section of the branch that's buried within the soil. This branch is thereafter sliced-off the parent plant, which has made the roots to develop and become new plants (Henderson, 2008).

Technically, it can be described as a method of flower propagation, whereby a portion of an above-soil stem develops roots while still connected to the parent crop, but then detaches later as an independent crop. Layering has grown from a common technique of vegetative propagation to now covering numerous species within the natural environments. It's also commonly utilized by floriculturists to develop desirable plants. For natural layering, it occurs when the branch comes into contact with the ground, thereby producing adventitious roots. For a much later phase, attachment with the parent crop is severed whereupon a new crop is generated as a result.

The floricultural layering procedure typically involves lacerating the target part, in order to depict or showcase the inner stem, or applying rooting compounds which is an optional measure. As for ground layering, the ornament plant's stem is bent downwards and the identified region buried inside the soil. This is performed in plant nurseries, which imitate natural layering by various plants like brambles that bow over and contact the tip on the soil, at which place it grows roots then when separated, may continue as a different plant. Whatever the case, this rooting process could take anywhere from a few weeks to an entire year.

Overall, layering is bit more complex compared cuttings, though has an advantage whereby the propagated part can continue receiving water and nutrients coming from the parent crop while it's developing roots. This is central for floras that develop roots gradually, or for propagating big pieces (Henderson, 2008).

Types of layering:

- Ground layering: Ground layering is essentially the preferred propagation method for many floriculturists. Original plants are put in the ground where the stem lies nearly horizontal, this forces side buds to germinate upward. Once these have started, the original stalk is buried in the ground up to its tip. During the growing season's end, the side branches shall have rooted, plus they can be divided while the crop is still dormant. A few of these shall be applied for grafting stocks, while some may be reused within the nursery during the next flower growing season.
- Air layering: For air layering, also known as marcotting, the plant's target region is cut and further surrounded in a humidity-retaining wrapper like sphagnum moss, this is further covered in a moisture cover like polyethylene film. A rooting hormone may further be applied to foster growth in the wounded parts. Once sufficient roots have developed from the wound, next to the stem cut from the parent crop is extracted and planted.The success of this germination method is determined by the availability of humidity, sufficient oxygen supply, and adequate temperature levels.

4.3.3 Grafting

Grafting is a method that vegetatively fuses two flowering plants into one. Rather than cross-pollinating two flowering plants and creating a hybrid seed, the grafted crops utilize the roots found and the lowest section of one plant, attaching it to a tender scion from the upper portion of another crop. This is frequently done with ornamental shrubs, where the best features of two plants are combined.

A graft hybrid crop thus carries with it a blend of desirable qualities comprising of new shoots, these are removed from a plant known as "graft variety." Equally, the root is availed by a crop called "pattern" or "rootstocks", which provides special radicular features to the graft. Currently, Japan, Spain and Korea are the focal producers of grafted crops in the world, each delivering 100s of millions of plants every year (Henderson, 2008).

Grafting is a technique used in flower plant nurseries to rapidly grow new crops. By the period the crop reaches you, the gardening center or store where you purchase the plant from, it may be impossible to know whether the crop is grafted or otherwise. The graft union would have since healed up making it hard to detect.

Bench Grafting:Bench grafting, also referred to as whip or tongue grafting is typically done indoors, preferably within a cool zone. A barn or unheated garage would make for ideal locations. This method is highly successful for grafting matter pencil-sized to around 1-inch in diameter. Due to the large amounts of cambial contact, this method heals faster and makes a rather strong union. Preferably, the stock and cutting must be the same width. This variety of graft is prepared when the scion and stock are dormant, usually from late March up to early April.

The scions must be collected just from latent terminal twigs. The scion plant comes from the preceding season's growth. In case you need to keep the scion for a short period, ensure it's stored in humid sphagnum moss, sandy, or a synthetic bag in a cool area. Scion wood should be kept humid and cool until usage. After the incisions are made, scions need to be inserted straightway. Never store scion wood inside a refrigerator or storeroom that has kept or still currently holds apples. They give off ethylene fumes, which can destroy the scion branch.Stock isusually prepared by placing a 1- to 2-inch slanting cut at the highest point of the flower stock. A second downward incision is created into the stock, beginning roughly 1/3 of the way below from the top side.

Cleft Graft:Also called cleft drafting, it is most commonly applied for top-working a fairly mature plant over to another form. The system is typically used on stocks ranging from 1-to 3 inches in width. Scions are prepared by doing a tapering cut measuring 1-to 2 inches long onto each side. Equally, the scion must be 3-to 4 inches long, plus just about 3/8 to half an inch in width. The stock is typically cut off squarely, then split vertically using a knife or forked grafting tool with a depth of around 2-to 3 inches.

4.3.4 Budding

Budding is a form of grafting whereby just the bud is applied rather than a portion of the scion with multiple buds. There are 2 types of budding which are; T-budding or shield and chip budding. Usually, budding is done during late summer, even though dormant chip budding may be done during late winter right before growth begins. Budding is the principal method applied by nursery workers growing ornamental plants.

A third variation whereby buds are propelled to grow in the same season of graft is known as June budding. Just as the name suggests, it's carried out during the planting season of late May up to early June. The June budding may either be a small chip or T-bud method. The top section of the rootstock usually is broken over, though not completely off, a few inches just above the bud around 3-to 4 weeks following the bud's insertion. As soon as bud growing starts, the highest of the root-stock is sliced. June budding shall only work in zones where there's a long, tepid growing season (Henderson, 2008).

Shield or T-Budding: T-budding activities are limited to the season of the summer whereby the bark starts to slip or separates effortlessly from the plant or wood. It usually happens from late May up to early June including from mid-July up to early September, though most budding is performed during the final period. Bud-sticks of a particular flower scion variety are derived from the mid-section of the present season's growth. Equally, the rootstock may either a 1yr old seedling or a rooted layer. It's also known as fall or dormant budding, which is among the most common budding systems used in floriculture.

In dormant budding, usually, the bud stays as a substitute on the rootstock's trunk until the subsequent spring, whereby the rootstock is chopped off right above the flower scion bud. A shrub may be produced within one year or less through June budding method, but creating a finished flowering plant through Dormant budding will require two growing seasons at least.Whichever the case, these scion buds are sliced from the budstick before placement in the "T" cut section. The bud shall then be wrapped into position using a rubber-bud strip. This rubber bud strip shall disintegrate within a few weeks. But if it doesn't, or you apply a non-biodegradable tie, then you'll have to slice the rubber band. There havebeen substantial success budding flowers using a crossed "T," but the only variance in the process is placement of a cross-cut at the base of the stock's vertical cut (Henderson, 2008).

Chip Budding: Chip budding is a procedure that can be applied whenever grown buds are present, plus the bark isn't slipping. This process can be performed in the spring right before growth starts plus when the stock, including the scion, are totally dormant. The chip budding process is typically used when stocks and scions are relatively small, usually an inch in diameter. Over the recent years, many commercial plant nurseries have gone from T-budding over to chip budding since the buds seem to grow out much better.

4.4 RAISING FLOWER CROP THROUGH TISSUE CULTURE

Tissue culture is a biological method of raising flower crops, whereby fragments of tissue coming from a plant are relocated to an artificial setup, from where they can continue surviving and functioning. The cultured tissue can comprise of one cell, a number of cells, or an entire or portion of an organ. The cells in culture might multiply; shift size, form, or utility; exhibit specialized function; or interrelate with other cells (Greer, 2007).

4.4.1 Culture Process

Preparation of flower tissue for the tissue culture procedure is done in aseptic conditions, with HEPA filtered atmosphere provided through a laminar-flow cabinet. Subsequently, the tissue is developed in sterile containers, like flasks or Petri dishes in a growth room having controlled temperature levels and light intensity. Additionally, living plant matter from the surroundings isnaturally spread on their surfaces with microbes, such that their surfaces are well sterilized in chemical content solutions, like alcohol or sodium or hypochlorite, before appropriate samples also called explants are taken out. These sterile explants are consequently placed on the top surface of an antiseptic solid culture medium, though they could at times be put straight into a sterilized liquid medium, especially when cell-suspension cultures are preferred.

Solid and liquid matter are generally comprised of inorganic salts, including a few organic vitamins, nutrients, and plant hormones. Compact media are primed from liquid media, typically with the adding of some gelling agents, such as purified agar. The context of the medium, especially the crop hormones and nitrogen source have significant effects on the form of the tissues which develop from the original explant. Case in point, addition of auxin may often cause a multiplying of roots, whereas a surplus of cytokinin could yield shoots (Greer, 2007).

The balance between auxin and cytokinin shall often generate an un-organized development of cells, commonly known as callus, though the form of the outgrowth shall depend on the crop species and the medium composition. When cultures are developed, single pieces are usually cut off and sub-cultured onto new media, this allows for growth or alteration of the culture's morphology. The ability and experience of a tissue culturist are central in determining which parts to culture, as well as those to discard. Once shoots appear from a cultured part, they can be sliced off or rooted using auxin in order to create plantlets that, upon maturity, can be transmitted to potting soil to foster increased growth in the green-house as typical plants.

4.4.2 Established Cell Lines and Main Cultures

There are two key types of cultures: which are primary or mortal cultures including cultures made from established or immortal cell lines. The primary cultures comprise of regular plant cells and tissues, removed directly from samples collected through biopsy from a living plant. Primary cultures are valuable in that they fundamentally model the natural properties of the material under study.

Nevertheless, the longer samples are preserved in a culture, the greater mutations they amass, which may cause alterations in chromosome structure including cell function. Additionally, primary cultures largely are mortal. Cells pass through an aging process in which they reproduce for around 50 to 100 generations, thereafter the rate decreases remarkably. The point in which cells found in primary cultures cease growing, or experience replicative senescence, points to the so-called Hayflick boundary. A biological replication limit discovered by U.S. microbiologist Leonard Hayflick (USDA, 2019).

In contrast, established cell lines may be perpetuated unlimitedly. These cell lines normally are derived from growth biopsies of plants or could be produced from primary cells which have undergone mutations, enabling them to surpass the Hayflick limit then continue replicating. Same to cells found in primary cultures, those in established lines gather mutations over time which can alter their character. Therefore, in order for scientists from varying laboratories to effortlessly compare results coming from experiments applying similar cell lines, they should first check the identity of those cells which they're working with. Generally, cell identity is determined through a process called authentication, whereby the DNA structure of cultured cells is contrasted against the recognized or standard profile that particular cell-line carries. The cells could be raised in a culture medium with biological origin like tissue extracts, in a chemically controlled synthetic model, or a mixture of these two. The medium should contain proper ratios of the required nutrients for the cells under study, plus be suitably acidic or alkaline. Cultures are typically grown either in form of single-cell layers or from a plastic surface, including suspension within a liquid or semi-solid medium. When producing culture, sometimes they are derived from single cells, which results in the creation of uniform biological copies called clones (USDA, 2019).

5

CULTIVATING FLOWER CROPS IN GREENHOUSES

CONTENTS

5.1 Introduction........ 112
5.2 Advantages Of Growing Flower Crops in Greenhouses........ 113
5.3 History Of Growing Flower Crops in Greenhouses 114
5.4 Growing Systems For Flower Crops in A Greenhouse 116
5.5 Characteristics Of Flower Growing Medium in a Greenhouse: 124
5.6 Important Flower Crops Commercially Grown in Greenhouses........ 133

5.1 INTRODUCTION

A greenhouse is a building made of the roof and with walls. It is also referred to as a glasshouse. The first site of modern greenhouses was in Italy during the 13th century. In the past, greenhouses were mainly found in rich homes but they later spread to the colleges. The largest greenhouses were built during the 19th century and it became popular since then.

The roof of a greenhouse is made of transparent materials that include glass and polythene. Plants such as flower crops are grown in greenhouses. Materials used in greenhouses construction are typically plastic or glass to allow the passage of light. Greenhouses are framed using steel, aluminum or woods like cypress and cedar. The sunlight brings warmth in the greenhouses by heating the ground inside. The plants are heated up due to the warming of air in the greenhouse. They are mainly used by farmers, botanists, flower growers, vegetable growers, fruits, tobacco and commercial plant growers. The functions of the greenhouses are to protect crop from excess heat or cold and pests.

Figure 5.1: Victoria amazonica in a large greenhouse at the Saint Petersburg botanical garden.

Source:https://en.m.wikipedia.org/wiki/Greenhouse

Greenhouses are mainly found in high altitude areas. The greenhouse allows gardeners to grow crops and flowers all season long. Glass greenhouses are used mainly for the production of vegetables and flowers. They have equipment for lighting, cooling, heating and screening. It can also be controlled by computers for optimum conditions for the plants (FAO, 2013).

5.2 ADVANTAGES OF GROWING FLOWER CROPS IN GREENHOUSES

The advantages of flower crops grown in a greenhouse are;

- The flower crops can be cultivated in any climate condition. They can be grown out of their season at a time in which they cannot be grown outdoor. Flower crops are protected from being damaged by wind, and rain, therefore, their market price rises. The flower crops are of high price in the market when they are out of the season.
- Flower crops are protected from many diseases and common pests. This is mainly to the soil-borne diseases that are splash by the rain to the plant. Pests such as rodents are put away by use of screens and traps.
- Production of flower crops in greenhouses minimizes the use of pesticides. When growing in a greenhouse, one can grow plants without the use of toxic pesticides.
- Greenhouses help in saving energy. Sources of energy like water can be controlled as compared to ancient farming.
- There is a huge good quality production. Greenhouses prevent flower crops from being stressed therefore improving production. The stress is caused by extreme temperature and high humidity.

Figure 5.2: Greenhouse production.

Source:https://www.novagric.com/en/blog/articles/advantages-greenhouse-production

- In greenhouses, there is an optimum use of technologies. This helps in climate management (Heating, humidity and saving energy). The technology used in greenhouses is of great importance to the growth of plants. Mainly in areas that have extreme weather conditions. Greenhouses also help in utilization of a land that has not be effective.
- Creation of optimum growing environment- gardener is able to create an optimum growing environment by regulating humidity, temperature and air in the greenhouse. This leads to a healthier growth of plants. In addition, better production is enhanced.
- A gardener customizes his or her needs. This means that the gardener can choose a greenhouse that is suitable forthe flower crop he or she wants to grow.
- One is able to grow many types of flower crops. An individual can experiment with a plant in the greenhouse that is not within the area. Various flower crops can be grown when using a greenhouse.
- The greenhouse keeps insects that have benefits - Insects such as the lady bags help in controlling dangerous insects in a greenhouse. The ladybugs can easily leave in open air.
- In greenhouses, there are plastic mulches. They help in filtering radiation coming from the sun. Greenhouse traps the thermal energy from the sun rays.
- Greenhouse leads to high income. Gardeners that practice greenhouse gardening gets high yields which they sell and earn a lot of cash. The crops are always big in size in greenhouse farming.
- Growing flower crops in the greenhouse lead to employment. Young people who have tried venture into greenhouse farming have earned lots of profits. This can help in solving the rate of unemployment.

5.3 HISTORY OF GROWING FLOWER CROPS IN GREENHOUSES

The ideas of growing crops in greenhouses started during Roman times. Farmers from Roman practiced artificial methods of growing cucumber that resembles the system of the greenhouse. Wheel carts were used for planting cucumbers that were put in the sun daily. They were then returned inside at night to get warmth. The frames were then used for cucumber storage. They were also stored in houses with sheets or oiled cloths referred to as specularis (Greer, 2007).

Figure 5.3: The royal greenhouses of Laeken, a masterpiece of the 19th century.

Source:https://www.newworldencyclopedia.org/entry/Greenhouse

The greenhouse concept was also in the Netherlands and England during the 17th century. Some of the greenhouse attempts needed much labor during the night. Balancing or providing enough heat was a big problem. In 1681 the Chelsea physic garden in United Kingdom was the first heated greenhouse. Currently, Netherlands have large greenhouses that can produce up to a million vegetables.

In the 1800s, the first modern greenhouse was built in Holland by a botanist from France Charles Lucien Bonaparte. The greenhouse was used to grow medicinal plants. Only the rich had the greenhouses but later spread to the universities. This was due to the growth in the study of botany.

The French first greenhouses were referred to as orangeries because they protected the Oranges grown. Pineapples greenhouses were referred to as pineries.During the 17th century, there were upgraded constructions and better glasses in greenhouses in Europe.

The largest greenhouses were constructed in England. This was when the botanist and the wealth were in a competition of building more elaborate greenhouses. A good instance is in the creation of large greenhouses by experimenting glass and iron by Joseph Jaxton. In 1880 the first greenhouse was built in Japan by a British Samuel Cocking. During the 20th Century, there was the addition of a thin hemispherical shell to various greenhouses.

During the 1960s when polythene was widely spread, greenhouses structures made of polythene were adapted. Growers made hoop houses frequently and they were also made by companies. Greenhouse wasconstructed by aluminum and steel poles which reduced the cost of production. As a result, more greenhouses were built in small gardens.

The polythene became more durable when UV-inhibitors made and added in the 1970s. This led to the extension of life of the polythene use from two to four years.In the 1980s and 1990s there were gutter connected greenhouses that were more prevalent. There was decrease in heating input.

Today's greenhouses are becoming automated. They allow people to save time during watering. There are automated machines that control humidity and temperature. Greenhouses also have automated ventilation to regulate temperatures and humidity when it is needed. Making greenhouses has become cheaper due to the use of materials such as polycarbonate panels and fiberglass.

However, there is no change in the passion for gardening since greenhouses were first built. All gardeners still take care of the flowers in the greenhouse every time.

5.4 GROWING SYSTEMS FOR FLOWER CROPS IN A GREENHOUSE

5.4.1 Ground Beds

A proper foundation preparation is important when preparing the land. This helps in long term growth. It also helps in healthy plant growth. Labor for digging, removing existing plants is needed to set up the garden. It may also include the addition of soil inputs and preparation of seedbeds. This is done to support crop growth. Irrigation lines are added to supply water to the crops. Also mulched pathways and row covers are added. Next maintainability steps are then taken (FAO, 2013).

Ground gardening needs proper preparation and good planning.

- The flower crop to be grown is considered: The size of the land to be used depends on the type of flower crop that the farmer wants to grow. A ground garden for growing flowers is different from then one used in growing vegetables.
- The gardening area is considered: Slightly raised lands are suitable since there is no much water retention.
- Preparation of the gardening area: Pulling out unwanted and existing plants. Using a tiller to prepare the soil.
- The garden is tilled and the soil is tested: This is done to determine the nutrients in the soil.
- Do the plotting of the garden: This is done before planting. Place pots and containers where you need them to be. Leave spaces between plants to enhance proper growth.
- After planting you can start mulching the garden: Mulches prevent weeds from growing.

Ground beds system involves the direct growing of flower crops into the ground. It has so many advantages including;

- It is easy to irrigate. The system of irrigation is easy to start, design and set up. It is also easy to install compared to raised beds.
- Little water is required in-ground beds. This is because ground beds don't dry out fast hence it needs little water to be maintained.
- It requires little capital. Money is saved by not using imported soil but using the existing soil. The money can be used to upgrade the existing soil instead of purchasing topsoil. Purchased soil may be similar to the existing soil or manmade. When the existing soil is improved properly, it maybe of good quality more than that of manmade soil.
- Ground beds are not permanent. A crop can be replaced whenever needed.
- The starting labor is less. Since a flat area can be plowed.

Figure5.4: Ground bed preparation.

Source:https://www.pinterest.com.au/pin/394065036116094763/?amp_client_id=CLIENT_ID%28_%29&mweb_unauth_id=&simplified=true

Ground beds have some disadvantages that include;

- There is little drainage. Therefore, they are exposed to flooding. This will rely on weather condition.
- Compaction of soil in-ground beds. There may be a lot of clay soil or people may be walking frequently in the garden. This leads to compaction of soil. It also causes unhealthy crop growth.
- Early weeds are still found. The soil may have many seeds of weed. This can be controlled by practicing mulching.
- It takes time for the plant to soften out of freezing.

5.4.2 Raised Beds

Raise beds are constructed with wood or other building materials. The size of the raised beds depends on the materials used or the preference of the farmer. Gardens are raised at about 6-8 inches. Bricks and blocks can also be used to raise them. Things to consider when or before doing raised beds gardening are;

- They can be framed or unframed
- The site of the raised beds should have about 8 hours of sunlight for high plants produce
- The topsoil should be compost to fill up the raised beds.
- There must be a drip and simple irrigation of a raised bed garden. Soil heating increases the loss of moisture, therefore, requiring frequent irrigation during hot seasons.

There are also some advantages of raised beds system in flower cultivation. These include;

- It is easy to manage. It can be used in intensive farming of a small portion.
- A raised bed allows for the long growing season. They have better drainage and heat up so fast. This is if the soil is tilled properly. There is also good growing condition.
- There is little maintenance of the raised beds. There is no soil compaction when the soil is in raised beds. Therefore, minimum tilling seasonally. When raised beds are mulched properly and taken care of, weeds are controlled.

Figure 5.5: Raised bed gardening.

Source:https://eartheasy.com/farmstead-raised-garden-bed-three-sizes/

- The quality of soil is improved easily. There is a conservation of building materials in raised beds. There are proper management and good control of fertilizers, water, mulch and other farm inputs. Therefore, this results in less waste.
- People with disabilities can easily access raised beds. Farmers that have difficulties in bending can access it. When raised beds are of good height even those on wheelchairs can access them.
- Raised beds drain well. As compared to ground beds they have better drainage. This allows the growth of crops that would not have been grown in areas with poor drainage to grow.

- Preventing damages on soil and compactness of the soil: The structures used help in the protection of plants. The soil does not get compacted because there are walking paths. Therefore, people do not walk on raised beds and plants are not destroyed.

Raised beds also have some disadvantages;

- Raised beds require a wall construction: It requires a lot of labor to build the walls but recycled materials are used to build it. Raised beds are expensive. They require techniques in supporting heavy soil weight. Soil is required to be filled hence it becomes expensive.
- Raised beds do not favor the growth of some flower crops.
- Raised beds gardens require more labor for weeding, fertilizing the garden and planting crops.

5.4.3 Benches

There are many advantages of greenhouse benches as compared to the use of pots or direct planting of plants in the greenhouse soil. These include;

- There is a good display of plants.
- Advanced air circulation and control of environmental conditions are provided by the benches. Circulation of air is vital for good growth of plant. There is control of water since plants dry up more frequently. Therefore, fertilizer can be applied regarding the amount and time. Air circulation enhances rapid drying of plants' surface after watering hence preventing moisture. This leads to minimizing foliage and root diseases. The plants have a better growth since roots of the plants become warmer as compared to ground beds which have cooler roots.
- There is better control of diseases and pests as compared to other gardening systems.
- Flower crops are of good level for comfort work. This allows for better observation and eases maintenance without the need for bending to work. This applies majorly to the flower crops that have short heights.

Figure5.6: Ebb and flow type benches.

Source: https://www.zzhfmachinery.com/products/ebb-flow-roll-bench-ebb-flow-rolling-benches-ebb-and-flow-roll-bench/

The size of the bench depends on some factors that include;

- The height of the person operating on the benches: A tall gardener can operate in wide and high benches in comparison to a short gardener. An average-sized benche are good since it favors people of different heights.
- The types of crops that are grown in the benches,
- Ability to reach both sides of the benches.

When setting up a bench for flower crops growing, the spacing ratio of the benches is considered.

- Bench design- there are different bench designs that include; flat benches thatinvolve placing of plants on a table and step benches which plants are displayed well in a greenhouse. The disadvantage of step bench is that there is sunlight from only one direction towards the plant. As compared to other types of benches, step benches need more labor to build.

The following are other types of bench construction

- Corrugated materials benches - corrugated benches last longer since the material used is strong but it may get destroyed. Corrugated materials do not rot.
- Non corrugated materials benches - These benches are plastic made. They are not strong. These materials start to deteriorate with time.
- Wooden benches - Woods are used in making benches. If not taken care of they are not permanent and deteriorate with time.
- Wire mesh benches - Wire mesh benches control diseases and enhance good drainage.

5.4.4 Potted Flower Crops in a Greenhouse

Pots add appeal to the soil in the greenhouse garden. They can be used by farmers with small spaces or gardens. There is addition of versatility in both small and large gardens.

Figure5.7: Container flower gardening using pot.

Source: https://gardenerspath.com/how-to/containers/tips-for-beautiful-garden-containers/

For the maximum production of potted flower plants, there should be proper planning of the equipment and management. Greenhouses of about 100 feet are suitable for potted flower plants due to easy regulation of temperature.

In pot benches construction, wood and tile can be used. The potting height should be checked to enhance efficiency. The standard and the pan types are mainly used pots. They vary from 1 inch to 14 inches. Most of the efficient pots are clay made and they have a bottom hole for proper drainage. In the last stages of flower production, wooden pots are used since they appeal to the retailers. The air impervious pots have been said to be more efficient. The use of concrete pots is limited. A pot ought to be light in weight, it should be strong and averagely porous.

Potting and shifting: Potting is planting a seedling of big plants in a pot. Shifting refers to the moving of a plant from one pot to a larger sized pot. There are plants that require strong potting such as geraniums, while there are those that require less packed soil like Begonias.

Every pot should have the same water space. It is advisable to always deep and sprinkles new pots in water before being used. The new pot absorbs lots of moisture and water is used by the potted plants. Optimum temperature is needed by potted flowers. If the temperature is not maintained it will lead to unhealthy crop growth. There should be regulation of temperature. Soft succulent growth is caused by high temperatures. There should be thermometers placed in the greenhouse. Each point of the greenhouse should be checked since the temperatures may vary.

Management of the pots: The pot management depends on good bench spacing and crop rotation. It also depends on the temperature regulation ability, light, water and air regulation abilities. The temperature should be regulated all the time.

Good timing of flower crops is considered. Timing depends on the weather conditions.

Advantages of potted flower plants

- Flower crops that grow for a longer season can be grown in pots. The crops should get enough sunlight.
- Potting is cost effective- It is ease to acquire containers without time-wasting or much effort.
- Plotting is flexible-A gardener can arrange the flower crops to suit a certain season.

- Potted flower crops are mobile. Gardeners can easily move the plants around the greenhouse. This means you can move your garden from place to place since it is in the pot.

5.5 CHARACTERISTICS OF FLOWER GROWING MEDIUM IN A GREENHOUSE:

5.5.1 Provide Adequate Nutrients to the Flower Crop

For greenhouse plants to germinate and develop flowers, they require a stable source of essential minerals and nutrients, such as calcium, manganese, nitrogen, zinc, phosphorus and potassium. Crops intake these compounds from the soil when their roots absorb them from water dissolved in the soil. A plant's capacity to absorb nutrients is typically affected by the soil's pH level, including nutrient deficiencies or imbalances which may lead to problems for some flowering crops. Ornamental herbs have similar nutrients requirements as bedding crops early during their crop cycle. But, as the herb crops steadily advance from vegetative development into flowering, the fertility rates and nutritional ratios may change for each plant (Author, 2017).

Figure5.8: The greenhouse provides a safe and controlled environment for flower growing unlike in field planting.

Source: https://www.carlyleobserver.com/news/local-news/carlyle-s-country-garden-greenhouse-celebrates-15th-year-1.1933525

Essential nutrients: The 6 main nutrients vital to healthy and productive plant development are phosphorus, potassium, carbon, hydrogen, nitrogen and oxygen. The gases are commonly found in the atmosphere and water, but physical compounds always come from the earth. Other major elements that plants require in huge amounts are magnesium, calcium and sulfur. The essential compounds promote chlorophyll formation and the procedure of photosynthesis and protecting developing crops from environmental obstacles and disease. Additionally, the elements are needed for the healthy growthof flowers, roots and leaves; potassium particularly promotes vibrant flowering in developing plants.

Measurements: During the winter season, garden plants and herbs need less moisture than in summer as a result of reduced transpiration ratio. However, nutrient application amounts remain similar all-year-round as nourishment needs of a plant don't change much right from summer up to winter. This is mainly because the greenhouse assists growers to be less reliant on the external environment, where temperature, day length, light and precipitation often fluctuate. Since crops use less water during the winter plus nutritional requirements don't change, they get feed at slightly greater rates than during the summer such that they don't run out of nourishment between water applications (Author, 2017).

Typically, fertilizer levels are measured in terms of EC (dS/m). For instance, the EC of a fertilizer provided to roses during the final harvest season in spring is around 2.5 dS/m, with an optional leaching ratio of 10 to 30 percent per watering. It's also advisable that the leachate must have an EC which is one unit greater than the application EC of the fertilizer mix. It's better to examine the EC and overall pH of the solution input and output each day and do adjustments as required. Furthermore, these tests can validate the accurate operation of injectors or the particular nutrient concentration.

Signs that your crop is nutrient deficient: If flowering plants aren't thriving or generating flowers as expected, despite getting sufficient sun and water, plus aren't infected by illness or insect pests, then a nutrient imbalance or deficiency could be the cause. Nutritional deficiency or imbalance is identified by general signs like yellowing of leaves; the death of plant material; and buildup of anthocyanin, which causes purple or reddish crop parts; stunted crop growth and lack of fresh growth. Equally, for flowering plants, nutrient shortage may show as a deficiency of buds or improperly developed blossoms.

5.5.2 Support or Anchorage for the Flower Crops Grown

Anchorage can safeguard your crops from falling and suffering unnecessary damage. Fewer fallen flower containers translate to less broken branches, damaged flowers or spilled soil. You'll also be doing less clean up in your garden since the soil-pots won't be frequently toppling over.

There are several techniques that can be applied to stabilize your potted plants, preventing them from tilting over, even when accidentally knocked down by someone, or in case of high winds as is the case with semi-closed greenhouse structures. Small plants are generally more vulnerable than larger crops and shrubs, though they're also simpler to protect from cold, wind, or heavy rain that may sometimes hit partially covered greenhouses (Author, 2017).

Here are some effective ways you can anchor your flowers as a grower:

- Fill up the bottom of a lightweight container, preferably made from foam, fiberglass or plastic, with 1-or 2 inches of gravels or pebbles. This must be done before filling the bottle with soil. These rocks and pebbles strongly hold down the pot in order to prevent it from tilting over without compromising on water drainage. Basically, apply a 3-inch coating of organic mulch on top of the plant's root area to shield its roots from chilly weather coming on the backdrop of a storm. Ensure the mulch is around 3-inches from the crop's foliage. Apart from synthetic containers, other items that can be used are overturn planter pots, glass cloches or buckets, laundry baskets and large bowls which provide a physical barrier from the elements. For an even stronger hold, weigh down the casing with a cement slab or large rocks if you anticipate high winds.
- Alternatively, fit a smaller pot within a larger one then fill the area between these two containers with rocks or pebbles. You may also use a broader or longer planter plus arrange various pots inside, filling up the space using soil. Next, arrange pots against the fence, deck rail or any other structure which can assist in keeping them upstanding in heavy winds. More so you can keep them next to some windbreak structure, like a compact fence or

fixed against a wall. These pots are also not likely to slant over if you put them at the corner, or cluster them together. Additionally, consider wrapping larger vines, plants, and shrubs using some strong fabric thatcan help them endure wind or cold. Preferably use burlap or a floating row covering. Construct a simple frame by beating in 4 garden stakes about the crop if necessary. Ideally, tie on the fabric using some heavy twine.

- Other protective tips: If the storm is too much, consider moving vessel plants including any hanging baskets into your house, or garage. Sometimes, semi-permanent greenhouses may not fully protect from storm damage and transferring the plants to a safer zone may be the only viable option. However, if a potted crop is too big or heavy to transfer, protect it using a casing, in the same manner, you would safeguard a similar crop growing in soil. Still, don't risk physical injury on yourself by seeking to protect crops in the middle of a heavy hurricane or tornado.

5.5.3 Good Moisture Holding Capacity for the Flower Crops

One important thing healthy soils provide to crops is humidity. Soils with the capacity to hold lots of moisture for the flowering plants are highly valued, whereas those lacking the water-holding capability are considered less valuable. A soil having good texture and form will absorb a massive deal of rain-fall. The method of water drenching into the earth is called infiltration. After entering the soil, this water travels downward. This transfer of water via the soil is referred to as percolation. A soil which allows water movement through infiltration or percolation is considered to be permeable.

During heavy rainfall, the pores found in soil fill up faster than water is capable of draining through the soil. Once all the pores are full of water, the soil is regarded as saturated. Additional rainwater develops puddles or moves downhill above the soil surface as runoff. Usually, water accounts for around 25 percent of the soil's composition. The fraction is higher after a downpour but lower in dry periods. A soil's ability to retain water is known as moisture-holding potential (Author, 2017).

Forms of soil moisture: Water present in soil comes in 3 types which are: capillary moisture, gravitational moisture, or hygroscopic moisture.

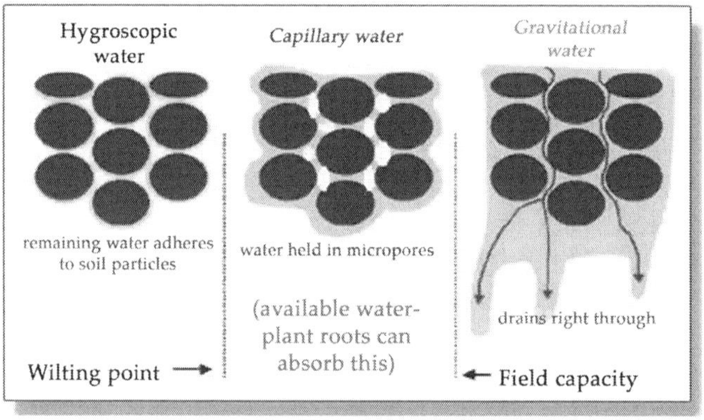

Figure5.9: Soil can be categorized based on its water-holding properties.

Source: https://slideplayer.com/slide/12098349/

- Gravitational moisture: Following a downpour, most of the water seeps down from the pore spaces found on the soil. It's a procedure called gravitational moisture since gravitational force pulls down the water. This rain ends up as underground water lying beneath the soil surface. Typically, it flows fast through sandy soils with large pore spaces but slower in clay soils containing tiny pore spaces. Plus, it's readily available for flowering plants to use. As this moisture travels through the soil, it bears with it dissolved chemicals, minerals, and salts in a process referred to as leaching. Overall soil texture will define how fast minerals, salts and chemicals are leached up, or washed, over soil. Due to leaching, crops germinated on sandy soils will require more regular fertilizer applications.
- Capillary moisture: Capillary moisture is the second form of soil water. It's also the water which crops are most capable of utilizing. It's basically the water held in-between soil particles against gravity's force. The moisture can travel upward or sideways via the soil through capillary action. When surface soil dries up, water slowly pushes up into the soil's open pore spaces through capillary action. Clay soils tend to have a greater pore spaces compared to sandy soils; thus, they contain more capillary water which is accessible to plants.
- Hygroscopic moisture: In this type, water develops a thin film which surrounds individual soil particles. The driest soils, also,

have hygroscopic humidity potential. Crops are unable to intake and use this form of water since it's held very tightly onto the soil particles.Basically, plant moisture can be classified as Available or Unavailable. The first option is when moisture found in the soil is consumable to plants, whereas Unavailable Water is present in a condition that's not beneficial to the crop. This property is usually determined by soil-moisture tension, referring to the force in which soil particles stick on to moisture.

5.5.4 Sufficiently Porous for Flower Crops

A porous soil consists of coarse particles thatleave gaps called pores in-between the spaces. The points between the clay particles, silt or sand that constitute the makeup of your soil is known as porosity. These further facilitate the air and water movement throughout the soil. The overall size and amount of pores present in your soil will affect the amount of water it can uphold, including how fast water moves from the soil. Porous loamy soils have the richest balance of all, blended with organic matter that retains water and delivers the nutrients required by crops. Clay and sandy soils typically have less organic material and encounter drainage problems: sand is extremely porous while clay is impermeable (Author, 2017).

Soil properties: Large soil particles, including those that clump together usually leave a lesser number of gaps, which are responsible for moving air and water through the molecules. Clay particles, in general, have the smallest porosity measurement with an overall diameter of 0.002mm. On the contrary, silt particles have a range of between 0.002 to 0.05 mm which is the ideal measurement, whereas sand is whichever molecule that's bigger than silt but lesser than 2 mm in terms of diameter.

Tips for making clay more porous: Fine clay particles merge together and inhibit water, air and crop roots from passing through the soil. An easy solution to the problem is mixing in organic matter, like compost or manure that provides essential nutrients for soil microbes. The waste emanating from these microorganisms makes the ground more porous since it forms spaces in-between the soil specks. You can buy bagged organic matter from garden supply centers to help with this task. For application, follow the steps below:

Apply a 4-inch deposit of compost or dry manure using a shovel upon the soil before growing. Cultivate the soil and mix in the organic material until you get a 9-inch coating of tilled soil. If you're adding organic content to an existing permanent flowering plant or shrub, apply a coating of organic material but don't till the soil.

- Add some mulch after planting your ornamental crops or flowers. The mulch sustains soil moisture and provides organic material over an extended duration as it gradually decomposes. Many gardening experts recommend pesticide-free grass clippings, these decompose quickly, particularly for annual flower plant beds, including bark mulch, which also disintegrates gradually, for perennial beds.
- Sow cover plant seeds during the fall season after harvesting your warm-season plants. During late winter or initial spring, mix the cover plants into the soil right before planting your regular crops. Some conducive cover plants include clover, alfalfa and fine grasses, all which contain leaves and elongated taproots that increase air pockets on the soil while they decay, meanwhile allowing air, water, nutrients and crop roots to pass easily across the porous soil. Additionally, put a 2-inch coating of organic material before the subsequent annual flower or ornamental crop growing. The soil requires enhancement from organic material before each annual growing.

Adequate soil porosity allows for oxygen to enter the root system, where it fosters growth.

5.5.5 Withstand Pasteurization with Steam or Solarization

Pasteurization is a bacteria elimination process, whereby water and some packaged and non-packaged items are treated using mild heat, typically to ratios of around 100 °C for purposes of eliminating pathogens. Spoilage enzymes found in seeds can also be inactivated using pasteurization. Nevertheless, since it's a chemical process it may have certain adverse effects. A better alternative to this procedure is soil solarization, which is natural and doesn't result inany side effects.

Soil solarization makes use of the sun's natural heat, trapped under transparent plastic sheeting, to manage many different types of harmful bacteria, fungi and nematodes. The procedure is typically performed in summer plus works best in areas with hot, sunny conditions for about 1 to 2 months straight; whereas daytime conditions above 80F degrees are ideal. Nevertheless, solarization isn't quite effective in coastal zones with summer fog; neither does it function well in extremely windy areas (FAO, 2013).

Here are some steps to guide you in solarization:

- Cultivate the soil, removing any weeds, debris, or large chunks of earth. It's crucial to remove all growing weeds since clear plastic — different from black plastic — does not stop the growth of crops in the soil underneath it.
- Prepare a bed measuring around 2 1/2 ft. wide, note that narrower beds don't allow for the easy buildup of heat plus it won't have as much effect on the germinating seeds. Proceed by carving out a tiny ditch around the perimeter, then rake it up to level surface. Next, soak soil to an average depth of 1ft: humid soil conduct heat much better compared to dry soil plus initiate sprouting of any potential weed seeds, which shall immediately be killed by the heat.
- Cover up the soil with around 1- to 4-mil transparent plastic; use UV-resistant synthetic material if it's available, considering it won't disintegrate during solarization. Stretch up the plastic tightly so it's in close contact with the earth. Proceed to bury the edges within the perimeter ditch. A secondary coating of plastic will increase heat levels, making your solarization procedure more effective; soda cans can be applied as spacers between these two sheets.
- Leave the plastic around for another 4 to 6 weeks before removing it. However, don't leave it below for much longer than 2 months, since the soil structure might suffer. You are now free to plant. Following the soil laying, avoid cultivating portions more than the topmost 2 inches of soil, considering weed seeds at much deeper levels might still be viable.

Soil solarization was first developed as a way of controlling soilborne pathogens. Later, it was discovered to be effective against various other soil borne vermin and weeds. It works by heating up the soil to appropriate temperatures of between 40–55°C for adequate durations of time. The procedure kills weed seeds found in the soil, including some perennial weeds comprising of perennating organs found near the surface. This is achieved through ensuring the soil is sustained at field capacity, and covering the surface with a transparent polythene cover. The soil must always be kept moist since this enhances heat conduction and thermal temperatures.

5.5.6 Free from Weed Seeds, Nematodes, etc.

Weeds act as alternative hosts for crop-parasitic nematodes, plus have long been known for their capacity to sustain nematode populations. Therefore, by eradicating weeds from your garden you'll also be controlling nematode numbers. Besides acting as alternative hosts, some weeds can further protect nematodes from chemical pesticides and other environmental dangers or contribute to fluctuations in future nematode biological potentials (FAO, 2013).

Weed removal: Weeds are nature's restorative remedy for venues that are in a damaged, plant-less state, however weeds and gardeners usually have different notions of what constitutes a good recovery. With a greater understanding of weeds, you can control them better and give your garden sufficient time to heal.

Sometimes, digging and cultivating only brings concealed weed seeds to the top, just imagine weed seeds are already there waiting to erupt, similar to ants from a disturbed anthill, each time you expose a patch of earth. Dig just when you have to and readily salve the troubled spot with crops or mulch. In grasslands, minimize soil disturbance through using a sharpened knife consisting of a narrow blade for slicing through the weed roots. This will sever their source of feed rather than completely digging them out. Have in mind that the weed seeds can stay dormant for a considerably long period of time (FAO, 2013).

Apply mulch: Mulch benefits crops by keeping the earth cool and humid and denying weeds of light. Natural mulches, in particular, may actually host insects like carabid beetles or crickets, which actively seek out and consume 1000s of weed seeds.

Some light can pass through dense mulches, and oftentimes you'll discover when it's too late, that particular mulch you applied was riddled with weed seeds. It is essential to replenish your mulch regularly to sustain it at around 2 inches deep, anything more than 3-inches deep will only deprive the soil of air. In whichever case, you can put weeds way back through covering up the soil's surface using a light-blocking piece of cardboard, biodegradable fabric or newspaper, then applying more mulch over it.

If you prefer using this method on rarely-dug areas, like the root-zones of flowers and shrubs, opt for solid landscape material for the light-blocking lowest sheet. There's a catch, nonetheless: When enough organic material collects on the landscape material, weed seeds that are dropped by flying birds or transported in on the wind shall begin to grow. As for the bottom part of the fabric, for it to be effective, the weeds should be pulled just before sinking their root systems into the ground. Additionally, for managing nematodes, plant cool-season flowers rather than warm-season ones. Nematodes are not quite active in cooler months, so there's less probability that they'll damage plants. Grow nematode-resistant crops all year long in order to limit damage. Generally, nematode resistance is specified on the plant or seed label. Likewise, amend the soil with sufficient organic matter before planting. Dig up or till the organic material a few inches into the earth. The organic matter shall help subdue the nematodes plus keep them from doing any major damage (FAO, 2013).

5.6 IMPORTANT FLOWER CROPS COMMERCIALLY GROWN IN GREENHOUSES

5.6.1 Rose

Figure5.10: The rose flower is popularly grown in greenhouses for aesthetic purposes.

Source: https://www.agrifarming.in/rose-cultivation

Roses are a form of flowering shrub. The name is derived from the Latin word Rosa. Typically, the flowers of a rose grow in different colors, ranging from the common red or yellow roses, to sometimes white and even purple roses. They belong to the plant family called Rosaceae. Most wild roses comprise of 5 petals and sepals. Larger amounts of petals are due to mutation which can be found in cultivated roses. These crops are also well renowned for their prickles. Each color of the rose provides a unique meaning: red, which is the lover's rose, symbolizes enduring passion; white, show humility and purity; yellow, friendship and joy; while pink roses represent gratitude, appreciation and adoration. The numberof stems found in a rose flower bouquet may also express particular sentiments. Usually, roses are grown as ornamental plants for their radiant flowers, either in the garden or sometimes indoors. They're also commonly used for perfumery and commercial grade flower crops (Plasmeijer and Yanai, 2012).

5.6.2 Carnation

Carnation is a double-flowered plant and usually a cultivated variety of the Old World pink, also known as Dianthus caryophyllus, which is available in a variety of colors. The most common blend is moderate red Carnation, that's also called grenadine or clove pink and is local to the Mediterranean region. This crop is widely grown for its fringe-shaped flowers, which usually have a spicy scent.The carnation flower is historically rich, with its scientific label interpreted to "flower of gods" or "flower of love," based on the source. It dates back to early Greek and Roman periods when it was applied in art and decorations. Carnations during these early periods were mostly found in varieties of peach and pale pink colors, however over the years palettes of available colors have grown to include yellow, white, red, purple, and also green. To date, carnations still remain a favorite floral choice for various occasions. These flowers are easy to recognize and possess a natural charm and allure which captivates most people (Plasmeijer and Yanai, 2012).

5.6.3 Chrysanthemum

Chrysanthemum, is a flowering genus comprising of around 40 different species of flowering crops within in Asteraceae family, its native predominantly to sub-tropical and temperate climates of the Middle East and East Asia. They are still common in these areas, where they're usually depicted in artworks. Cultivated forms, often referred to as mums, grow as fall-blooming plant ornamentals and are crucial in the flower industry. Florists' chrysanthemum comprises of well over 100 different cultivars, including pompon, daisy, button, and spider forms.Most crops found in this genus are categorized as perennial herbs or subshrubs. They have modest aromatic leaves thatline up along the stem. There are some with disk and ray shaped heads, while others lack this distinctive feature. Also, cultivated species and plant hybrids typically have larger flower heads; whereas those from wild species have smaller sizes. Still the taxonomy of this genus is complex and has undergone several revisions, now including costmary, tansy and feverfew flowers among others.

Figure5.11: Chrysanthemum flower in a greenhouse.

Source: https://www.pinterest.com/pin/146155950381360762/

5.6.4 Lilium

Lilium is a group of herbaceous flowering crops that grow from bulbs, and have large prominent flowers. The species are native to temperate northern climates, even though their range goes further into the northern subtropics. Lilies are tall perennial plants varying in height from around 2 to 6ft. They develop bare or tunicless scaly sub-ground bulbs, constituting their perennation organs. For some North American species, their bases develop into rhizomes, whereby numerous small bulbs can be found. There are also species thatdevelop stolons (Plasmeijer and Yanai, 2012).

These flowers are large, usually fragrant, and available in a variety of colors such as orange, pink, white, yellow, red and purple. Some may also have markings, such as brush strokes and spots. Lilium plants flower mostly during late spring and summer. Their flowers consist of umbels or racemes at the top of the stem, often with 6 tepals either spreading or reflexed, which give floras varying shapes ranging from a funnel to a "Turk's cap".

5.6.5 Gerbera

The Gerbera flower family is easily identified by its head-like inflorescences which resemble solitary floras but are usually composed of 10s to 100s of flowers which are often elaborate in structure and form. There are 3 main flower types thatare distinguished under this category; these are marginal ray flowers, central disk flowers and intermediate trans-flowers. The existence of different flower forms within one genotype makes gerbera quite a unique target for floricultural and evolutional studies, considering that the traditional model crops for flower development have only single floral types in their inflorescences.

More often, Gerbera flowers just have five petals. Their dorsal petals are narrow and slim in their ray and trans-florets, whereas ventral and lateral petals are merged together to create a ligule-like structure. The result is a bilaterally symmetrical figure of these flowers, commonly known as zygomorphy. For the central disc flowers, usually petals are shorter and not well-fused up, plus the center-most floret is radially symmetrical having separate petals (Plasmeijer and Yanai, 2012).

5.6.6 Tulip

Tulips are popularly utilized as wedding-flowers in table garlands, bridal bouquets or even general decoration. Likewise, when used for regular gardening purposes, tulips have the potential to liven up space through their array of beautiful shades. Moreover, some tulip bulbs are actually edible and can be used during cooking.The tulip's flowers, on the other hand, are large-sized and radially symmetric with some containing both male and female characteristics. Normally erect, or sometimes pendulous, these are arranged more typically as single terminal flowers, or if pluriflor then as two, three up to four petals. For structure, the blossom is normally cup or star formed. Plus, like with other flowers in the Liliaceae group, they are undifferentiated and bi-seriate formed from three petals and sepals, though sometimes these are called tepals since they are almost identical. These tepals are usually petal-like in form, and brightly colored thus making them ideal for decorative purposes as well.

Figure 5.12: Tulip flowers in a greenhouse.

Source: https://www.floraldaily.com/article/9074609/new-greenhouse-complex-for-tulip-and-greenery-growing-opened-in-smolensk/

5.6.7 Anthurium

Anthurium is a class of herbs that often grow as epiphytes upon other crops. They are both terrestrial and waterborne depending on the species. Their leaves are usually clustered and variable in shape, plus the inflorescence bears tiny flowers thatare perfect, having male and female parts. These flowers are enclosed in dense spirals found on the spadix. Plus, the spadix is usually extended into a spike form, though it may also be globe or club-shaped. Underneath the spadix is a spathe, which is a form of bract.

It's variable in shape, also, though lance-shaped in several species. The Anthurium may stretch out flat or as a curve. At times, it may cover the spadix in the form of a hood. For Anthurium breeders, the spathe and spadix form the main focus, where they are often grown in bright colors and spectacular shapes. A. andraeanum and Anthurium scherzerianum, which are two of the widely common species in cultivation, form the only plants that develop bright red spathes (Plasmeijer and Yanai, 2012).

5.6.8 Orchids

The Orchid or Orchidaceae family consists of around 1,000 genera including well over 22,000 species of elaborately flowered plants spread throughout the world, particularly in wet tropics. This crop belongs to the order of monocotyledonous flower plants, which also comprises the iris and asparagus families. In nature, there are infinite variations of shape and shades in their flowers. Moreover, because genetic inconsistencies are fragile among orchid genera, horticulturalists are able to make successful crosses among different species in the genus, plus cross between the genera as well. There are over 22,000 species of plants in the orchid family, making it second in scope only to the Asteraceae family.

Structural diversity: Orchids are slightly different from other crops not just in morphological characteristics, but flower organization also. They have masses of pollen known as pollinia, fused stamens and pistil that form a column, plus most are herbaceous or non-woody in structure. Orchid flowers further vary immensely in size, ranging from 2 millimeters to 38 centimeters or more.

5.6.9 Alstroemeria

Alstroemeria, commonly referred to as the Incas Lily, Peruvian Lily or Parrot Lily originates from South America and comprises of a genus with around 50 types of flowering plants, mostly from the cool, hilly territories in the Andes.This crop is slightly zygomorphic or bilaterally symmetrical, consisting of 3 sepals and another 3, ordinarily, striped petals. The petals and sepals found on Alstroemeria plants are comparable in color and consistency, that is, there are no firm green sepals.

Alstroemeria contains 6 stamens and a completely undivided style. Equally, the ovary found on an Alstroemeria is usually inferior, having 3 carpels. They feature a monocot plan where floral parts exist in 3s. This group is largely like grass, whereby the veins moves up the leaves though don't branch across. If you look keenly at an Alstroemeria flower stem, you may sometimes see a helix growth pattern around the stem. It's due to the creation of new cells thatdevelop in a spiral arrangement.

5.6.10 Eustoma

Eustoma Flower, also called the Texas Bluebell or Lisianths, comes from the Greek terms 'eu' meaning beautiful and 'stoma' which is mouth. It defines the blossom as 'beautiful mouth' because of its coloring and profile. Eustoma, is typically called prairie gentian in some regions. The plant is native to warm environments in the southern U.S., Mexico, the Caribbean and north South America.This genre is commonly found in grasslands, including zones of disturbed ground. The Eustoma Flowers have long stems and can typically grow 1 to 3 feet tall, plus the petals develop an attractive trumpet shape. These plants are most often seen in grasslands, where they bloom in summer and have a rich blend of colors varying from pink, purples, lavenders, and double color combinations (Plasmeijer and Yanai, 2012).

Eustoma Flower is commonly used to signify a thoughtful, peaceful feeling; they can be applied to accent garden zones or as indoor floret arrangements. The flower is also quite low maintenance and capable of withstanding higher temperatures and moisture indoors.

5.6.11 Gypsophila

Gypsophila paniculata, also called panicled baby's-breath, is a genre of flowering crop found in the Caryophyllaceae group, where it's native to central and east Europe. The plant is a herbaceous perennial which grows to 4ft tall and wide, having mounds of dividing stems shielded in clouds of small white flowers during summer, thus the name "baby's breath." This flower's natural habitat is found on the prairies of dry, sandy and gravel areas as well as calcareous soils that look like chalk.

Figure 5.13: Gypsophila in a greenhouse.

Source: https://www.worldatlas.com/articles/global-leaders-in-cut-flower-exports.html

Gypsophila is a common ornamental garden plant, thrives mostly in well-drained moderate to alkaline soils in the full sun. Various cultivars have been developed from this flower species, of which 'Rosenschleier' and

'Bristol Fairy' have been identifiedamong the best. Equally, G. paniculata is greatly used in the flower business, where it's often referred to as "gyp." It provides a great backdrop for bigger or more defined blooms, with most of it commercially grown in Peru for exports (Plasmeijer and Yanai, 2012).

5.6.12 Statice

Statice flowers are vibrant Mediterranean imports thathave found great favor with crafters, florists, and home gardeners who value its showy, long-lived bracts. You may not find statice crops in many nurseries, though their seeds are affordable and sprout easily even for the brownish thumbs. Because of its funky, defined aroma, some would find it suitable for outdoor décor, where they complement seaside landscapes or rock gardens.

The other classic names of statice are seafoam, marsh-rosemary sea and lavender, which all provide a clue as to the nature of Limonium genus where it belongs. Statice crops thrive mostly in coastal grounds, shrugging off the salt-laden soil and ocean spray. As for height, they can reach 15 to-30 inches tall provided there's sufficient warm weather throughout the year. Complete sun exposure is a crucial aspect of growth, without which the plant may flop over. Statice crops come in a range of springtime colors, such as white, pink, rose, yellow, violet and peach shades.

6

CHALLENGES AND CONSTRAINTS IN FLOWER CROPS CULTIVATION

CONTENTS

6.1 Introduction 142

6.2 Deficiency Information on New Varieties of Flower Crops 142

6.3 Lack of Precise Information on The Production of Different Flower 143

6.4 Poor Transportation 144

6.5 Lack of Organized Marketing Co-Operatives 145

6.6 Lack of Research And Development In Floriculture 146

6.7 Lack of Market Intelligence 147

6.8 Lack of Detailed Economic Feasibility Studies 147

6.9 Lack of Streamlined Quality Control Mechanism 148

6.10 Poor Co-Ordination Between Government And Private Agencies Involved In Import And Export of Floriculture Products 149

6.11 Lack of Information on Advanced Cultivation Practices 150

6.12 Lack of Appropriate Planting Materials 150

6.13 Lack of Knowledge About Importing And Exporting Countries 151

6.14 Lack of Quality Consciousness of The Produce 152

6.15 Lack of Sale Promotion Activities 153

6.1 INTRODUCTION

Despite generating millions of dollars each year in revenue, the floriculture sector continues to face numerous challenges. Key among these setbacks is limited availability of capital, including minimal access to cheap credit for flower farmers. In fact, scientists estimate that the primary cause of poor productivity in the sector is insufficient credit to finance procurement of farming inputs including capital investment. Great interest rates have made it untenable for flower farmers to seek credit. Also, insufficient storage facilities for harvested products impede marketability of crops; some of them reach the market in poor quality, thus making them unattractive to buyers. Some farmers also have to contend with lack of processing amenities near the sources of flower production, which has greatly restrained the level of exploitation and potential for growth in this sub-sector. Furthermore, the prices of fertilizers including other agrochemicals are rising gradually, making them exorbitant to many flower farmers thus increasing the overall production per unit cost.

Moreover, underdeveloped rural roads including other key physical amenities have led to increased costs of moving horticultural produce to the markets, as well as farm inputs.This trend has continued to diminish the competitiveness of floricultural produce in many countries. Additionally, electricity in most rural areas is costly or even inaccessible, thereby affecting investment in facilities like cold storage, agro-processing and irrigation that are necessary for production.

6.2 DEFICIENCY INFORMATION ON NEW VARIETIES OF FLOWER CROPS

There are new varieties of plants thatcan be grown either as specialty cut-flowers or applied to provide some extra height to garden beds, specimen bowls and combination planters. These new varieties are more profitable for farmers and durable as well. Some of them are as follows:

6.2.1 Dianthus

It's a flower class comprising of several new strains. Bright Eyes of 'Amethyst' is a multi-bloomer which produces glaucous, azure green leaves, including scented large dual crisp white flowers having a bright maroon eye on solid stems. This new flower breed has a dense mounding habit, typically reaching 14 to -18 inches tall and 10 to-14 inches broad. Also found in this group is 'Waterloo Sunset', a successive bloomer, producing multitudes of fragranced double blooms supported on sturdy stems having glaucous, blue emerald foliage.

6.2.2 Lisianthus

Also known as EUSTOMA, this series is less susceptible to rosetting in plug-growing, plus less prone to brief stem length within long-day conditions. Crops have double flowers on robust stems. Furthermore, the colors making up this series are categorized by group, whereby group I are the earliest to bloom, and group IV the last. The F1 Vulcan flower series, in particular, has 6 colors. It's easily recognized by the thick, cup-shaped floral petals which enhance its transportability.

6.3 LACK OF PRECISE INFORMATION ON THE PRODUCTION OF DIFFERENT FLOWER

One of the most common errors beginner farmers do is picking the wrong plant, something that can lead to poor yields. A good understanding ofthe climatic conditions of your area, type of soil and flower breeds is necessary if you are to succeed in growing flowers. Basically, these plants can be classified into 3 categories depending on their blooming behavior. They include annuals, biennials and perennials.

Perennials are quite popular in landscaping projects. They have the longest lifespan around three years, though some may stay alive for even longer with proper care and environmental conditions. On the other hand, annuals last for only one season though they are more affordable compared to perennials.

Biennial flowers have a similar cycle though it will last for 2years instead of one. In the preliminary year, plants grow and stem, but won't bloom. For the second year, they will bloom for one season, and then die. Additionally, plants need sufficient amounts of sunlight to germinate. You should be cognizant of the amount of light your flowers receive per day, for instance, partial-shade plants require sunlight only for a small portion of the day, in the morning or evening. For full-sun plants, they can withstand midday heat and will make do with at least 7hours of sunlight to flourish (Dait, 2015).

6.4 POOR TRANSPORTATION

There are many bottlenecks and constraints related to the transportation of floriculture produce to the market. Case in point, seasonally blocked roads especially in rural areas and irregular transport services, alongside poor storage, may cause losses since flowers are perishable and deteriorate quickly with time.

Figure 6.1: A rural Columbian farmer transports his flower produce to market through poor roads that can flood after a downpour.

Source: https://slate.com/news-and-politics/2015/02/colombias-flower-industry-on-farms-outside-bogota-valentines-day-is-less-hearts-and-kisses-and-more-big-bucks-and-sore-backs.html

Transport opens upmarket for floricultural produce, improves networking among geographical and commercial zones plus introduces new regions to economic focus. Therefore, if transportation services are scarce, of low quality or costly then growers will be disadvantaged when seeking to sell their produce. If flowers are transported through rugged roads then they might also suffer damage from bruising, consequently resulting in lower rates for the cultivator. Infrastructural neglect of rural areas is especially rampant in developing countries, where farmers have to deal with poor roads year in year out. Sometimes, after heavy downpour or storms, these roads may become completely inaccessible making it impossible for people to access the market. The floriculture sector desperately needs organized road systems and smooth transport networks to improve growth, employment and even economic opportunities within communities. Even though roads are the most common means of transport, alternatives like rail and air transport are increasingly being used as well by flower farmers to transport their produce since they are highly convenient (Dait, 2015).

6.5 LACK OF ORGANIZED MARKETING CO-OPERATIVES

Marketing cooperatives are useful in shielding floriculturists from poor practices and misdemeanors by sellers and brokers. Normally, the management of these structures is done through market committees, with nominees from local bodies, brokers, farmers, state officials and other stakeholders in the industry. In the absence of such structures, growers will have to contend with unauthorized rate deductions, underhand dealings, or wrong practices by brokers in the entire market cycle. Additionally, in some countries, the cooperatives provide up-to-date and dependable market insight to the farmers. This is inclusive of the best flower breeds to grow in certain seasons, or those that will bring them the most returns for their investments (Thapa & Dhimal, 2017).

Organized marketing systems also provide grading and standardization within the industry, so that growers can package their produce accordingly based on quality. If this structure is lacking, it often leads to heap sales whereby all qualities of flower produce are sold within a single common lot. In such an environment, floriculturists are unable to enjoy the fair price for enhanced produce, and this means that there aren't any incentives for using improved farm inputs or producing better flower varieties. Considering the farmer producing quality yields isn't assured of any greater rates, but only the single standard rate cutting across all market products.

6.6 LACK OF RESEARCH AND DEVELOPMENT IN FLORICULTURE

Figure 6.2: Constant research is necessary for developing and improving flower varieties.

Source: https://en.wikipedia.org/wiki/Horticulture

Effective research, development, including training of farmers, is essential for meeting the global needs for floral products. Oftentimes, national governments use public resources under the agriculture ministry to conduct these studies.

Some countries have even engaged private sector institutions to support these efforts. The advantages of this public-private partnership are improved efficiency, including scaling of innovation, processes and outcomes within the floriculture industry. Innovation projects instigated through agronomic research and development programs usually have the capacity to impact beyond the preliminary target zone, or direct beneficiaries. Additionally, some researches seek to clarify important concepts and terminologies used in the floriculture sector, further providing a detailed assessment of farming approaches applied in the field, and drawing inferences as well. One key finding is that the average rate of return on investments within the floriculture sector is considerably high. Nevertheless, specific research outcomes still differ based on techniques used and modeling assumptions, especially models touching on research lag distribution, nature of research-induced scientific change, including the nature of markets for the concerned floral commodities (Thapa & Dhimal, 2017).

Moreover, through participatory approaches to detecting and analyzing emerging issues, including the designing and testing of innovations to combat planting challenges there's a high likelihood that farmers will enjoy improved yields through implementing research findings in the floriculture sector.

6.7 LACK OF MARKET INTELLIGENCE

Market data and intelligence are crucial in enabling flower farmers, as well as traders to make knowledgeable decisions around what flowers to grow, when to reap, to which markets products should be sent, including whether to store the crop or not. The most vital marketing intelligence needed by farmers, however, is price intelligence. Most growers today still lack adequate understanding and capability to use appropriate market intelligence, in determining their production or marketing decisions. Dissemination of market information, regarding demand, production and rates, plays a crucial role in the overall functioning of the entire market, by coordinating the competitive marketing procedures. Moreover, by helping ensure flower produce is transported to markets where there's actual demand for it, this shortens marketing channels plus significantly reduces transport costs. This further helps to ensure each marketing operation is done fairly, and all participants in the flower chain share in the business's risks and benefits. Modern advances in technology are also making it possible to provide flower farmers with the relevant marketing information they require. Nevertheless, farmers may sometimes not benefit from complex facilities, if the structure is ill-managed or not fashioned for their needs. It's not enough to avail market information, but it must also be distributed in a form thatis accessible to the farmers (Dait, 2015).

6.8 LACK OF DETAILED ECONOMIC FEASIBILITY STUDIES

Before starting a greenhouse, consider the size and facilities you will need to keep it running. Obviously, a bigger greenhouse will need more material, which shall ultimately increase the cost. Typically, the serious gardener shall make do with a 10x10 greenhouse as the minimum size. Though for the casual grower, 6x6 is typically fine. If you still aren't sure, go for a larger facility then fill it up slowly, since it's usually more complex to add-on to some already existing smaller structure.

A beginner's greenhouse is fairly low-cost, only a 6'x8' hoop-house would be ideal for getting started in moderate climates. There's no plumbing or fancy electronics required either. For more experienced growers, a larger 12'x12' greenhouse shall be more suitable. The materials applied will also have an impact on the cost of your greenhouse. Double strength glass is the most preferred option for siding material. It's not only beautiful but gives your building a permanent look as well. Maintenance involves cleaning up the same way you'd do to the windows within your home, though also inspect occasionally for chips, cracks, and breaks, particularly after a storm. For the frames, they are typically made of steel or wood which holds up the siding panels (Thapa & Dhimal, 2017).

6.9 LACK OF STREAMLINED QUALITY CONTROL MECHANISM

A coherent quality assurance (QA) structure, throughout the harvest and retail stages of flower growing, is essential in providing a top-quality supply of floricultural produce to consumers, plus protecting the reputation of particular marketing labels. Control mechanisms may begin with choosing a preferred genotype, then determining or recommending its appropriate harvesting time, preferred appearance, texture, and aromatic quality.

Careful harvesting, including handling, is necessary for minimizing physical injuries. Postharvest handling has the possibility to either preserve or reduce quality of flowers. Nowadays, the availability of affordable microcomputers, in addition to solid-state imaging processors has resulted in improved usage of computer-assisted video inspection to carefully sort products according to grade before marketing. Presently, objective and non-destructive means of differentiating floricultural perishables based on their flavor and structural quality are getting tested. Once approved, these shall become effective quality control tools thatare reliable and efficient (Thapa & Dhimal, 2017).

Safety assurance focuses on reducing chemical and microbial pollution during production, picking, and post-harvest handling of both intact and fresh-cut flowers. Going forward, research and development endeavors should concentrate on developing improved methods of assessing the quality and safety qualities of fresh products, being part of the quality control structure. Field inspectors should also ensure appropriate hygiene standards are met during the processing and packaging stages of products.

6.10 POOR CO-ORDINATION BETWEEN GOVERNMENT AND PRIVATE AGENCIES INVOLVED IN IMPORT AND EXPORT OF FLORICULTURE PRODUCTS

Flower industry stakeholders are in agreement concerning the need for collaboration in spearheading development in the sector. Towards this end, state and private agencies need to draft policies that promote standardization of floral imports and exports. Some of the goals to be met through these partnerships include:

- Promoting a stable and competitive production environment among member states.
- Encouraging participation of farmers in the trade cycle, and having them aware of emerging market affairs and trends,
- Strengthening flower business relations within regional and global marketplaces, and
- Achieving unrestricted movement of produce, capital, labor and services across countries

The statutory role of stakeholders in the import and export trade is promoting, developing and coordinating the production, including marketing of floricultural produce. All over the world, this sub-sector can be an efficient solution for different countries' needs for cash-crop diversification, income generation, employment opportunities and foreign exchange generation, in addition to availing raw products for agro-processing industries such as perfume making.

Additionally, at the point of entry, they may champion the availing of facilities like storage, warehouses and streamlined customs procedures to make it easier for produce to pass across. This way, both importers and exporters won't have to worry about their produce going stale during the processing stage (Thapa & Dhimal, 2017).

6.11 LACK OF INFORMATION ON ADVANCED CULTIVATION PRACTICES

Effective flower cultivation should seek to improve plant yield, quality, provide insect and disease resistance, and reduce environmental pollution. For this to be achieved, the farmer should have knowledge on modern growing techniques. Among the popular growing methods is the micro-irrigation system, commonly referred to as drip irrigation, which is a crucial strategy in curbing the effect of drought on germinating blossoms. This technique conserves water and you can irrigate nearly twice more land with it (Dait, 2015).

Farmers are also encouraged to use greenhouses; they protect against environmental hazards while providing a controlled setup for germinating crops. Polyhouse is another form of greenhouse that's slightly smaller in size, while polyethylene is typically used for the cover. In this method, everything from temperature, moisture, and fertilizer is strictly automated. Additionally, some floriculturists use cold chain storage methods to keep their produce looking fresher for longer. This is the process of controlling temperatures in both storage and transportation amenities, in order to prevent the rotting of perishable flowers. Usually, dry ice or solid CO_2 is introduced to the facility at temperatures of around -80°C. Cold chain storage is useful for keeping consignments frozen for extended durations of time, which is beneficial for farmers who transport their flowers long-distance to the market.

6.12 LACK OF APPROPRIATE PLANTING MATERIALS

Scientists in the flower growing sector are constantly testing and adjusting new technologies, mostly these are aimed at improving the productivity of flower growing throughout the world. Considering the complexity of this industry, modern technologies can help reduce constraints which affect the ability of small-scale farmers to attain high profitability levels with top-value flower harvests. Production technologies seek to overcome the challenges of land, capital, labor and infrastructure.

In some of the developing countries, ratios of post-harvest loss surpass 50%, plus cold storage is almost non-existent because of the high costs of cultivation equipment and erratic electricity supply. Rapidly cooling produce following harvest prolongs shelf life by minimizing metabolic activity, moisture loss and bacterial growth. Technically, farmers who can keep their produce longer have access to better prices, since market rates for flowers sometimes fluctuate. As for planting materials, drying beads can be used to provide a largely adaptable technique for drying seeds plus it's re-usable. When properly utilized, it often leads to greater germination rates including increased yield capacity. When integrated with local seed structures, drying beads can help improve the market for both local and enhanced cultivars. For tropical climates, it's also been identified that high humidity can cause quick seed deterioration, which results in reduced productivity and market value for farmers (Thapa & Dhimal, 2017).

6.13 LACK OF KNOWLEDGE ABOUT IMPORTING AND EXPORTING COUNTRIES

Figure 6.3: A flower expo in Poland showing quality flower strains shipped from different countries.

Source: http://www.sigma.com.pl/flower-expo-2016/?lang=en

All countries in the world and geographical zones have their own policies regarding the importation and exportation of flower products. It's important for flower farmers to learn about these policies, in order to increase the probability of their products being approved in aforeign country. Some jurisdictions have export quantity limits, defined as the overall physical quantity of local origin or cultivated products shipped outside of the country. This may further include re-exports.

For most regions, the value of floricultural exports is expressed in terms of US dollars, which is the globally recognized medium of exchange. However, you should still do your own research to determine what other forms of currency are accepted in the import/export country. As for import values, these are usually reported in form of CIF which stands for cost insurance and freight. It refers to the value of produce, including services provided to transport goods towards the exporting country's border.

Also, the value of services rendered towards delivering flower produce from the exporting country's border to the borderline of the importing nation should be considered when learning about a specific region's import/export factors. By learning about the flower business dynamics of a given country, flowers would be better placed to produce items thatmeet the jurisdiction's set standards of quality (Dait, 2015).

6.14 LACK OF QUALITY CONSCIOUSNESS OF THE PRODUCE

Quality product consciousness throughout the processes of harvesting, to the retail exhibition is essential in ensuring a consistently high-grade supply of garden-fresh floricultural crops to the clients, including safeguarding the repute of a particular marketing label. It starts with the assortment of the genotype, including its proper period to planting and reaping for the finest appearance, textural, flavor and health quality. Considerate harvesting and handling arerequired to prevent external damages on flower.

In fact, each post-reaping handling procedure has the capacity to maintain or reduce the value of a flower crop. For a few cases, such as growing of climacteric crops, they also improve scent quality. Nowadays, farmers use objective and non-destructive approaches of differentiating flowers produce perishables on foundation of their flavor or physical quality. Still, more research is being carried out to identify excellent tools for grading, preserving and handling of different flower breeds.

Some also seek to improve the shelf-life of produce, referring to the time duration in which a product maintains its sensory, functional, and physical characteristics within optimal conditions, and in a way that's acceptable for the final consumer. Numerous studies show that sensory variations are the key most important aspects which affect the quality of produce, showing the significance of using sensory evaluations in determining the shelf-life span of fresh flower cuts (Dait, 2015).

6.15 LACK OF SALE PROMOTION ACTIVITIES

Most floriculture promoters have come across the phrase that "effective marketing is key for profitability", however many still do not fully comprehend what it means. Mostly, the tendency is associating marketing with the process of "selling" or "advertising". They are, in fact, just a couple of facets of marketing. From an agribusiness perspective, marketing can be described as the overall system of trade activities intended to plan, value, promote, and dispense flower products that fulfill the needs of potential customers, whereas achieving the traders' objectives.

The main emphasis in this regard is the client, first and foremost. The trader that provides clients what they want, when including how they exactly want it, shall be the most successful in this sector. For a business as big and diverse as floriculture, it's impossible to create a single marketing program which would cover all participants in the industry since management skills, market prospects and resource available vary considerably from one region to another (Dait, 2015).

While creating a market-plan alone won't guarantee success, it still does insure that aspects affecting the viability and survivability of a flower business are considered. For a strategic marketing planning process, it should be practical, workable and address the market's needs for fresh and high-quality flower produce.

7

PESTS AND DISEASE AFFECTING FLOWER CROPS

CONTENTS

7.1 Introduction 156
7.2 Insects Causing Damage by Chewing 156
7.3 Insects Causing Damage by Piercing-Sucking 162
7.4 Mining Insects 166
7.5 Boring Insects 167
7.6 Major Diseases of Flower Crops 171
7.7 Management of Pests And Diseases of Flower Crops 183

7.1 INTRODUCTION

Pests which nest on flowers can chew on plant parts, tearing or cutting them into bits. They slowly remove tissue from the external margin inwards, plus in severe situations, most of the plant may be consumed, therefore completely destroying the germinating plant. Also, some pests carry pathogens that spread disease to the plant. Furthermore, among the common diseases is Botrytis, which usually occurs when conditions are appropriate. It's a fungal infection with the biggest host range, this disease is often called gray mold since it produces considerable amounts of fuzzy gray spores onto the surfaces of infected tissues. One common symptom is badly matted flowers.

7.2 INSECTS CAUSING DAMAGE BY CHEWING

7.2.1 Caterpillars

Figure 7.1: Caterpillars are moth and butterfly larvae that feed of foliage.

*Source:*https://www.shutterstock.com/video/clip-24819572-moth-larvae-caterpillars-green-color-eat-mock

Caterpillars are the larvae of several kinds of moths and butterflies which infest growing flower crops. One of the devastating ones is the variegated cut-worm, where larvae hatch from eggs brought forth in the spring season, and mostly feed after dark. Flowers may be sliced off at or around the ground overnight. Moreover, some breeds of cutworms have been known to climb and consume the foliage from above going downwards. One cutworm can destroy several crops in one night. Newly implanted annual flowers are the most vulnerable (Adil, 2016). Equally, other damaging caterpillar forms include;

- Yellow woollybear
- Beet armyworms
- Checkerspot butterfly
- Columbine skipper
- Diamond-back moth
- Painted lady-butterfly larvae

7.2.2 Flies (Maggots) and Midges

Figure 7.2: Transparent maggot pests feeding on a flower crop.

Source: http://www.pestnet.org/SummariesofMessages/Pests/PestsEntities/Insects/Aphids,PlanthoppersBugs/Aphids,Hibiscusrosasinensis,Germany.aspx

Maggots are larvae from flies, typically of the common house-fly or bluebottle types. Flies are mostly attracted to food including other rubbish; where they lay eggs on rubbish; and later these eggs hatch to form maggots. You will only have problems with maggots in case flies get to your poorly discarded waste, found around the flower growing site. The maggot stage of various flies can cause immense damage to cut flowers like Helichrysum and Chrysanthemums. For instance, the sunflower maggot is known to infest stems of the Helichrysum crop, and causing them to collapse. Also, the Chrysanthemum gall midge digs into leaves, stems and buds forming cone-shaped galls (Adil, 2016).

7.2.3 Beetles

Figure 7.3: A beetle chewing on fresh flower.

Source: http://naturesperspective.com/july-and-august-japanese-beetle-time/

Cut flowers are susceptible to attack from one or several species of beetles. In fact, both adult and larvalbeetles have chewing mouth-sections which potentially cause damage. Case in point, adult Japanese beetles consume a wide variety of flowering crops during the day causing massive destruction. Adults surface from the soil during the months of June and July, feeding on foliage for around 30-45 days. Right before the female dies; she delivers about 50 eggs beneath the soil surface in gardens or other grassy zones. Soon, these eggs develop into grubs which feed on flower and grassroots, slowly moving upwards as the soil warms up during spring (Adil, 2016).

7.2.4 Sawfly Larvae

Figure 7.4: Sawfly larvae hanging onto a flower for food.

Source: https://www.flickr.com/photos/orb9220/14273495063/

Sawfly larvae appear similar to caterpillars, though they're a totally different class of insects. They are related to wasps and bees, belonging to the order of Hymenoptera. These larvae typically feed on flower crop foliage, and can quickly lay to waste a rose garden within a short period. There are well over 8,000 types of sawflies, labeled so due to their saw-like appearance found in the female ovipositor, which is the organ used to input eggs in crop stems or leaves. Even though sawflies are closely interrelated with stinging insects, they don't sting themselves but rather consume pollen and nectar found in flowers. Their management may involve use of mild chemical sprays.

7.2.5 Grasshoppers

Figure 7.5: Grasshoppers have sharp incisors on their mandibles used to tear down crops.

Source: https://www.shutterstock.com/video/clip-8117410-green-grasshopper-on-rose-flower-garden-4k

Grasshoppers are mostly leaf and grass eaters; they can devastate cultivated flower crops when they move in swarms. Their populations also vary from one year to another;the damage is most likely to occur in years when arid weather accompanies population increment. Generally, drought conditions diminish natural vegetation, pushing grasshoppers to relocate to cultivated crops like flower gardens where they can get easy fodder. Usually, grasshoppers are tan to grayish-green in color, with considerably long and prominent heads including large compound eyes. The overwintering egg phase is usually nested in the soil. While nymphs hatch from eggs during the late May to early June periods. Both nymphs and adults of grasshoppers can feed on planted flowers (Adil, 2016).

7.2.6 Slugs and Snails

Figure 7.6: Slugs and snails may appear similar but snails have a shell for differentiation.

Source: https://pixabay.com/fr/photos/escargot-rose-p%C3%A9tales-coquille-2378220/

Snails and slugs consume on a range of flower varieties including decaying plant material. They create uneven holes with flat edges on flowers by rubbing with their raspy tongues. Tiny succulent crop parts are easily sheared off these creatures when they feed. The russet garden snail, commonly called Cornu aspersum, is among the most common species found in American gardens. The vermin was first introduced from France in the 1850s for consumption as food. Yet another destructive snail is Theba pisana, a white garden snail commonly found in California. Snails and slugs both are members of mollusk phylum, which appear similar biology and structure, only that slugs don't have external spiral shells like the snails.

7.3 INSECTS CAUSING DAMAGE BY PIERCING-SUCKING

7.3.1 Leafhoppers

Figure 7.7: A colorful leafhopper insect with serrated appendages.

Source: https://www.britannica.com/animal/leafhopper

Leafhoppers are ferocious tiny insects having an insatiable appetite. Their damage toflower plants can be widespread, so learning the best ways to kill leafhoppers in your garden or lawn is important. Damage from these bugs is marked by light-colored spotting on plant leaves, which results from leafhoppers sucking up sap and flower juices from inside the plant tissue. When left unchecked, this steady feeding reduces the crop's vigor over time, thereby browning the leaves. Destruction from leafhoppers is typically not intense enough to extremely harm mature plants; though, young plants or new buddings can be stunted or distorted by leafhopper feeding (Adil, 2016).

7.3.2 Thrips

Figure 7.8: Thrips can be identified by their yellowish-brown body and tough exoskeleton.

Source: https://sprayers101.com/thrips-flowers/

Thrips are small, slender bugs with elaborately fringed wings and distinctive asymmetrical mouthparts. Diverse thrips species forage mostly on floras by puncturing and slurping up the contents, while a few of them are predators. Around 6,000 species of thrips have been identified.Being a common pest in flower greenhouses and both indoor/ outdoor gardens, thrips may turn flowers and leaves pale, splotchy, then silvery, before they die. Damaged plants are twisted, stained and scarred. Similarly, a few thrip species acts as vectors for around 20 viruses which cause plant illnesses, such as the Tospoviruses. Few species, however, are useful as pollinators or predate on other insects.

7.3.3 Aphids

Figure 7.9: Aphids have a rounded-body and long slim legs.

Source: https://sophiespatch.com.au/2017/08/24/aphids/

Aphids are minute sap-sucking pests and members of Aphidoidea insect family, other common names are greenfly and blackfly, but individuals within particular species may vary broadly in color. The aphids are small pear-shaped, soft-bodied bugs that suck up the juices present in leaves, stems and flower plants. These pests can either be green, orange, red, brown, golden, grey, white or black in color. Mostly, they feed in compact clusters and don't move quickly when disturbed. Even though normally wingless, when these insects become too stressed or crowded, some shall develop wings and fly away to colonize new crops. Scientists estimate that around 4,000 aphid species are in existence(Adil, 2016).

7.3.4 Plant Bugs

The Miridae, also known as plant bugs, isa big and diverse insect order which at one time was known as Capsidae. This particular suborder, Heteroptera, includes well over 10,000 different species, while new ones are further being identified constantly. The most widely recognized mirids are those species which are notorious garden pests, they pierce flower tissues, feed on their sap, and sometimes spread viral crop diseases. A few species nonetheless are predatory.For their shape, plant bugs are oval-shaped or extended and measure less than 12mm in length. Most of them feature a hunched look, due to their prothorax shape that moves the head focused downwards(Adil, 2016).

7.3.5 Mites

Figure 7.10: Mites have a hairy body and a set of eight legs also covered with hair.

Source: https://www.hortweek.com/pest-disease-factsheet-spider-mites/plant-health/article/1438073

Mites can be found on ornamental, landscape, or houseplants. They are versatile tiny insects which can reduce your flowers' vigor, plus transmit various dangerous viruses and illnesses. Crop mites feed by biting through flower cells and suckling on the moisture found inside. The activity leaves behind necrotic and yellow spots on affected plants. Speckling on floral parts is a clear sign of mite presence. Among the easier to identify mite species are spider mites, which produce elaborate webbing that delicately wraps flower tips and stems. The spider mites attack various plants ranging from indoor-to-outdoor species. Mites are profuse breeders and populations may reach a harmful level very fast.

7.3.6 Whiteflies

The whitefly is a sap-feeding insect that's commonly found in dense crowds on the bottom of leaves or flowers. When affected plants are disturbed, huge clouds of winged adults hover into the air. These flies suck flower juices and, consequently, produce a gluey substance called honeydew. When honeydew is left by itself, it causes fungal illnesses to form on the crops. In the whiteflies are eating well, they'll produce numerous eggs on the bottom sides of leaves. Due to constant feeding, plants will rapidly become weak and might be unable to conduct photosynthesis. The leaves will also wilt, become pale or yellow, plus growth may be stunted(Adil, 2016).

7.4 MINING INSECTS

7.4.1 Leafminers

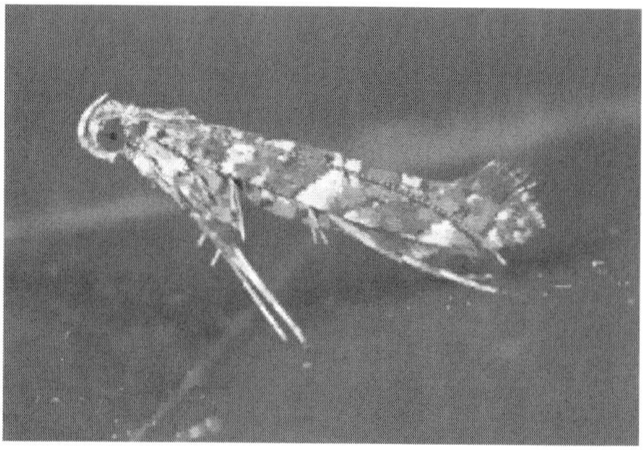

Figure 7.11: A leafminer sinking its mandibles into foliage, they usually create spiral blotches on leaves.

Source: http://termirepel.com/blog/?p=184

Leafminers consist of larvae from various insects such as beetles, moths and flies. They are typically very small in size, reaching about 1/5" to 1/8" in length. These larvae live within the leaves of crops and flowers, feeding on their soft, inner tissues found between the lower and upper margins of the leaf. When feeding, they usually produce a squiggly-like line on their leaves, which yields a mottled shape on the leaves. Even though leafminers seldom kill or rarely injure plants, they will deform and give it a rather unpleasant look. Once the larvae have finished feeding, they chew their way straight out of the flowers or leaves and fall to the earth to pupate.

7.5 BORING INSECTS

7.5.1 Stalk Borers

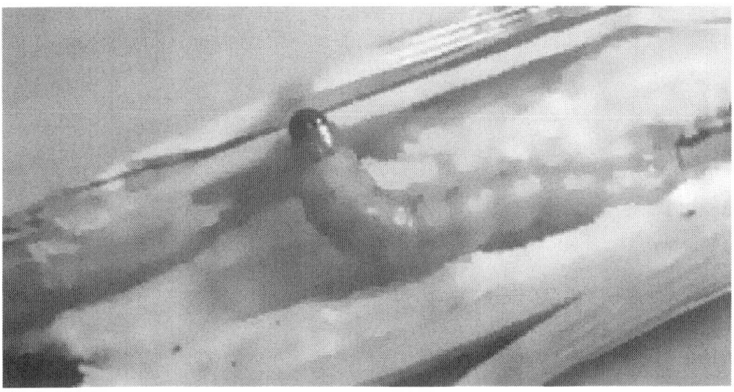

Figure 7.12: Stalk borers usually live in the stalks of infested plants.

Source: https://www.greenlife.co.ke/stalk-borer/

The stalk borer, also scientifically called Busseola fusca, is a common insect pest found mostly into East Africa. It a member of the Noctuidae family that also include cutworms and armyworms. But as the name suggests, the larva of stalk borers dig into the stalk or stem of host plants and feeds inside the plant.They lay eggs in a single column upon the stem beneath the leaf sheaths, where it hatches after around 10 days into a larval state. These mature in around 35 days thereafter, before pupating in the stem. These pupae persist for 2-3weeks then turn into adults. Usually, just one larva is present per stem since they are cannibalistic(Adil, 2016).

7.5.2 Burdock Borer

Figure 7.13: With its distinctive mosaic-patterned wing, the burdock borer moth is quite easy to detect.

Source: https://bugguide.net/node/view/156261

The burdock borer moth is scientifically known as Papaipema cataphracta, belonging in the moth family of Noctuidae. It's natively found in Canada, and the States of Florida and West Louisiana in the U.S. On average, it has a wingspan of 29–45 mm and flies from the months of August up to October based on the location.The larvae of owlet moths, which is their other name, dig into the roots or stems of their flower host plants where they often cause massive destruction. These insects are categorized by a metathorax structure known as the epaulette or nodular sclerite, which serves to keep parasites away from the tympanal cavity.

7.5.3 Iris Borer

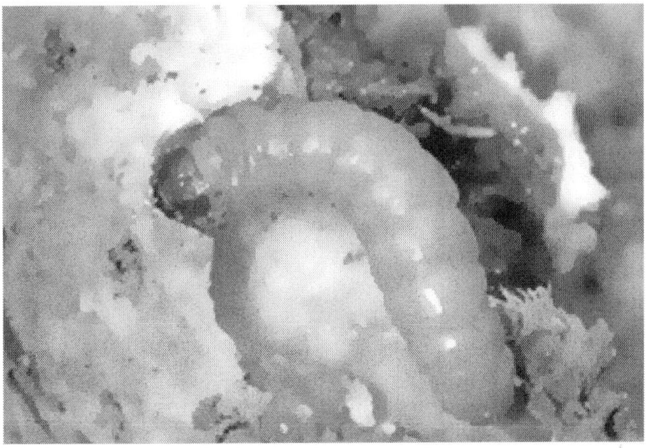

Figure 7.14: A pink colored iris borer destroying flowers.

Source: https://dengarden.com/gardening/How-to-Rid-Your-Garden-of-Iris-Borers

Iris borer caterpillars are scientifically known as Macronoctua onusta, and are among the most damaging insect pests within the iris subclass. These pinkish caterpillars grow to two inches long upon maturity. Generally, adult moths bear chocolate-brown frontal wings and much lighter yellow-brown rear wings having a wingspan of up to 2 inches. Their moths are seldom visible since they mostly fly at night. They burrow pinprick-sized spaces and tunnel within leaves while continuing down towards the rhizome, including the food-storage part for the crop which sends roots downwards whereas shoots having sword-shaped leaves move upward. Iris borer chewing causes water-soaked streaks which allow for bacterial entry(Adil, 2016).

7.5.4 European Corn Borer

Figure 7.15: A European corn borer with its triangular orange-brown wings.

Source: https://www.canr.msu.edu/ipm/diseases/european_corn_borer

European corn borer is also called the European high-flyer or European corn worm. It's a moth within the family of Crambidae that includes various other grass moths. It's a grain pest mostly, but it can sometimes affect other crops. The pest is local to Europe; it originally infests variations of millet, such as broom corn. This vermin was first recounted in North America during the year 1917 in Massachusetts, though it probably entered Europe a couple of years earlier. Following its initial discovery within the Americas, this insect later spread into Canada going further westward across the U.S. into the Rocky Mountains. The adult female can be identified by its yellowish-brown body(Adil, 2016).

7.6 MAJOR DISEASES OF FLOWER CROPS

7.6.1 Bacterial Diseases

Crown gall rot: The crown gall is a typical plant infection caused by soil-borne bacteria Agrobacterium tumefaciens. The bacteria which trigger crown gall can endure in the earth for several years and are released once galls get saturated with water or when older galls decompose. Young flower plants can easily be killed by maturing gall tissue.

Figure 7.16: Crown gall rot makes plants very unsightly to look at.

Source: https://www.gardenmanage.com/statuses/1000148755.html

This crop disease is caused by bacteria Agrobacterium tumefaciens, also called the Rhizobium radiobacter. Several thousands of crop species are susceptible, especially members of rose flower family, shrubs, vines, and the perennial garden plants. Some symptoms of infection include round, rough-surfaced galls which are several centimeters to a meter in diameter, typically near the soil perimeter (Everyday Science, 2019).

7.6.2 Fungal Diseases

Black spot: Blackspot is a fungi infection of flowers, which is usually problematic in foggy or moist coastal zones. It can happen anywhere leaves generally remain wet, like where flowers are sprinkler grown during the night or evening. The black spot fungal disease starts as black dots on the leaves, eventually causing them to become yellow and fall off. Apart from looking unsightly, it may seriously deteriorate the flowering plant. This pathogen endures on living or dead leaves including on infected stems. The spores are spread through splashing water, mostly from rainfall. Moreover, for spore making and infection, leaves should stay wet for around 7 hours or more.

Figure 7.17: Blackspot fungi causes dark blemishes on flower leaf surfaces.

Source: https://www.rhs.org.uk/advice/profile?PID=270

Powdery mildew:Powdery mildew infections are caused by various species of fungi found in the order of Erysiphales. It's also known as Podosphaera xanthii, growing mostly in environments having low moisture and moderate temperatures. The greenhouse provides an ideal humid, temperate climate for the spreading of this disease. Found on many crops and easily identifiable, the powdery mildew disease is found all across the U.S. It's caused by various closely related fungi species, each having a partial host range. Symptoms usually show up later in the maturing season on outdoor crops. The powdery mildew begins on young leaves as raised blister-like parts which cause them to curl, revealing their lower leaf surface.

Figure 7.18: Powdery mildew causes plants to appear frost-like.

Source: https://rightplantz.com/pest-problem-powdery-mildew/

Downy mildew: Downy mildew is scientifically called Plasmopara viticola. It's a foliage disease caused by the fungus-like organism known as Oomycete. The infection is spread from one plant to another through airborne spores. It's an infection of wet weather since infection is ideal in lengthy leaf wetness conditions; affected plants usually develop yellow to white spots on the upper layers of older leaves. This syndrome overwinters on crop debris and within the soil. Likewise, fungal spores may be transported by insects, rain, wind, or garden tools.

Figure 7.19: Downy mildew spreading across a garden.

Source: https://www.hortweek.com/pest-disease-factsheet-downy-mildew-affecting-ornamentals-crops/plant-health/article/1392737

For the undersides, they are comprised of white to gray, cotton-like fungi. The "downy" masses are usually noticed after the downpour or heavy dew, but disappear soon after the sunny weather resumes (Everyday Science, 2019).

Rust: Rust is a fungal plant disease, causing red, orange, copper, or purplish spots to show on the leaves. Once the disease progresses, leaves might eventually become crisp or brown and drop off even while the plant still has ample nutrition. Equally, infected leaves may curl up, twist and drop off from the crop. Controlling rust on flowers doesn't require costly chemical sprays thatpollute your garden or home.

Figure 7.20: Leaf rust is caused by fungi present in the environment.

Source: https://invasives.org.au/blog/major-biosecurity-and-pest-news-from-across-the-country/

There are natural methods of controlling rust on flower plants that can be applied. One of them is avoiding overhead watering and excess-watering, which favors growth and spread of fungi spores. Increased moisture and humidity also elevate the chances of rust infections.

Anthracnose: Anthracnose or Colletotrichum gloeosporioides is a disease that attacks all plant parts during the growth stage of flowers. Its symptoms are most noticeable on leaves and mature flowers. In the beginning, anthracnose usually appears on leaves in the form of small and uneven yellow, brown, or black spots(Adil, 2016).

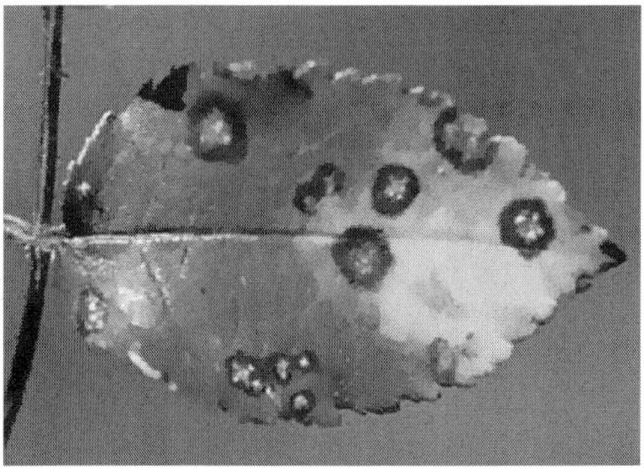

Figure 7.21: Anthracnose can make leaves lose their capacity to conduct photosynthesis.

Source: http://www.sactorose.org/ipm/83anthracnose.htm

Normally found in eastern sides of the U.S., anthracnose is triggered by fungi found in the genus of Colletotrichum, which is a common class of crop pathogens that cause diseases on various plant species. Infested crops develop dark, moisture soaked lesions on their stems, leaves or flowers. The cores of these lesions may further develop pink, gelatinous groups of spores.

Grey mold: Gray mold rot is marked by tan-to brown soft blotches thatbecome covered in a dusty mold. These can cause young shoots, seedlings and mature flowers to wither and flop. The flowers may also become speckled and rotten. This disease principally affects buds and flowers, even though contagion on leaves and stems is still possible(Adil, 2016).

Figure 7.22: Grey mold can form both on leaves and the soil surface.

Source: https://pnwhandbooks.org/node/3161/print

Control measures involve the total removal of infected sections. Appropriate cultivation, including sufficient spacing for ventilation, careful fertilizing, and water application at the soil-line, these are helpful in minimizing the probabilities of developing gray mold rot. Harvested flowers should further be stored at 32 °F in a freezer to prevent multiplication.

Verticillium wilt: Verticillium wilt infects a large portion of woody and herbaceous species. The causative fungus, Verticillium dahliae, contaminates susceptible plants via the roots. This disease has caused most destruction where, flower plants were cultivated in fields previously used for growing other susceptible crops like cotton, melons, tomato and peppers.

Figure 7.23: This disease makes crops to appear weak and dreary.

Source: https://www.amkhaseed.com/blogs/growing-information/tips-for-dealing-with-verticillium-wilt-in-your-garden

The fungi survive as tiny, black resting microbes capable of enduring in soils for multiple years. Once a susceptible crop is grown in a soil affected by fungus, the microsclerotia will develop and disease the plant. This insect invades and infiltrates the plant's vascular structure, plugging up the xylem and reducing the transportation of water across the roots system.

Sooty molds: Sooty mold is also called black mold or blotch. It's marked by splotchy black spots or coatings found on leaves, flowers and stems. The dark filtrate of sooty mold consists of dark fungi threads of a group of ascomycetes, as well as species of Cladosporium, Fumago, Alternaria, Capnodium and Scorias. The fungi grow in fluid sap or honeydew emitted by aphids including other sucking bugs.

Figure 7.24: Sooty molds often turn leaf surfaces from green to black in color.

Source: https://www.heartspm.com/pest-control-black-sooty-mold.php

While the condition is usually unsightly, it's not harmful. General control involves spraying or dusting the plant for sucking bugs, washing off sticky sap coatings or honeydew present on plant surfaces, including avoidance of any physical damage on the plants(Adil, 2016).

Canker: Canker is a bacterial syndrome that causes swellings on leaves, flowers and stems. It's not hazardous to humans but may result inpre-mature leaf and flower drop which eventually renders the plant unproductive. Flowers infested with canker may be safe to handle, though too unsightly to display in your home. Citrus canker might be spread quickly through wind or rain, landscaping gear, individual's hands or clothing, including by moving infected crops or plant parts. Bacterial canker results from two closely linked bacteria thatinfect the stems or leaves of flower species. They begin to form during mid-spring and then afterward shoots might die back.

Figure 7.25: Canker lesions devastating a piece of leaf.

Source: http://www.agriculture.gov.au/pests-diseases-weeds/plant/citrus-canker

7.6.3 Viruses and Virus-like Diseases

Rose mosaic: The rose mosaic syndrome is caused by the Prunus Necrotic Ring-spot Virus (PNRSV), including Apple Mosaic Virus (ApMV). They are common viruses found globally and known to cause severe illness in both fruits and flowers. Both viruses, separately and mutually, are found around rose bushes. In general, the mosaic disease causes flower parts to change color—typically to yellow. For someone with an artistic eye, this pattern may sometimes look like a tile mosaic, though this is very variable. For roses, the symptom is typically gaunt, splotchy yellowing around the leaves, though sometimes it could have on a whitish-yellow line on the flower leaf that circles around(Adil, 2016).

Figure 7.26: Rose mosaic disease affecting a germinating rose flower leaf.

Source: https://pnwhandbooks.org/plantdisease/host-disease/rose-rosa-spp-hybrids-rose-mosaic

Rose wilt: Rose wilt is a virus complex commonly known as "dieback" within some areas. This disease may be spread through vectors like aphids, plus symptoms are quite variable and vary from inhibited growth to curly young leaves. Soft tissue signs are more evident during spring and fresh leaves will impulse towards their specific petioles. Typically, affected leaves are fragile and easily drop from the plant. Wholly formed leaves shall appear to wilt like the crop were water-stressed, even though it receives enough moisture. The illness doesn't kill the flowering plant outright, though, over time, loss of leaves may weaken the crop making it susceptible to different stresses including winter damage.

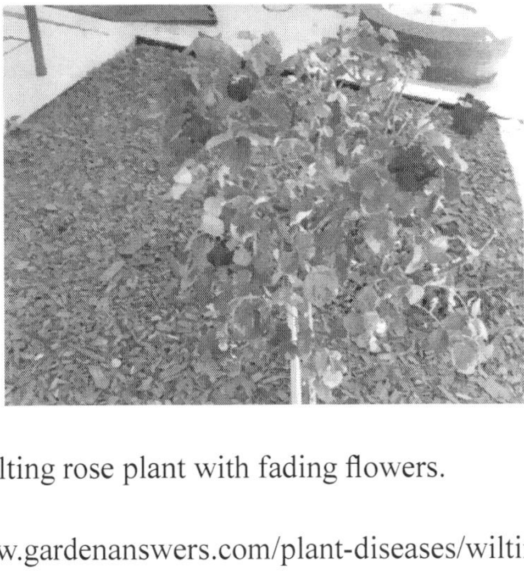

Figure 7.27: A wilting rose plant with fading flowers.

Source: http://www.gardenanswers.com/plant-diseases/wilting-rose-bush/

Rose rosette disease: This infection is caused by the newly discovered virus, called rose rosette virus. It's usually transmitted through an eriophyid or rose-leaf curl mite that inhabits shoot tips from sprouting, and leaf petal bottoms of roses. The disease spreads through grafting, but not seed or any other typical vectors. Initially, the rose rosette was mistaken to be phytoplasma disease; though, heat and tetracycline remedies didn't cure the infection indicating that a phytoplasma isn't the causal agent. Also called the 'witches' broom, it's fatal in the flower class of Rosa multiflora, where its average lifespan following infection is 22 months. The plant normally grows wild, though it could appear as hedges in some homes(Adil, 2016).

Figure 7.28: Rose rosette eats up flowers making them fade away.

Source: https://homesteadgardens.com/rose-rosette-disease-is-bad-news-for-rose-lovers/

7.6.4 Environmental Disorders

Frost: Frost damage on crops occurs when water contained within the flowering plant, specifically individual plant cells, get frozen. As the water gets cold, it develops ice crystals inside the plant which actually puncture crop cell walls. The leaves and tender fresh growth are normally affected first. Originally, they will look wilted. Thereafter the wilted growth shall turn black or brown and ultimately become crispy.

Figure 7.29: Frost can destroy flowers by freezing their water content.

*Source:*https://www.gardeningknowhow.com/ornamental/flowers/roses/roses-winter-care.htm

The crops can get damaged or killed through freezing temperatures or actual frost. Effects vary depending on the kind of plant, specific tissue exposed, including how long low temperatures are present. The "light frost" starts from −2 to 0 °C which damages fewer kinds of flowers, while "hard frost" is below −2 °C.

Salinity: Salts present in the earth's water may impede plant growth for various reasons. First, the existence of salt within the soil solution lessens the ability of a plant to intake water, and this causes a decline in the growth potential. The process is known as osmotic or water-deficiency effect on salinity. One telltale sign of salinity in flowering plants is stunted growth, where plant leaves usually have a bluish-green tint. When salt levels within the soil surge to more lethal ratios, scalding or burning of the tip and peripheries of the much older leaves occurs. Eventually, the leaf expires and falls off then finally, the plant perishes.

Figure 7.30: Excess salinity can affect the growth of roses.

Source: https://homeguides.sfgate.com/salt-affect-growth-roses-31946.html

Herbicide damage: Herbicide damage refers to any adverse, mechanical damage on a flowering plant caused by contact of that plant with pesticide intended for weed control. The symptoms of impairment may vary based on the chemical used and specific crop affected. Some common signs include yellowing, leaf curling or cupping, odd leaf texture, dead or necrotic cells and reddening of veins. Some crops are very delicate to herbicides, whereas others may be more resistant towards certain types of herbicides. Any flower can be affected bythis problem. Usually, pigment inhibitors in some herbicide chemicals disintegrate chlorophyll, which is required for photosynthesis to happen. They also interfere with germination and growth.

Figure 7.31: Glyphosphate is one of the dangerous chemicals found in herbicides. It can deform flowers.

Source: https://content.ces.ncsu.edu/glyphosate

7.7 MANAGEMENT OF PESTS AND DISEASES OF FLOWER CROPS

7.7.1 Scouting and Early Detection

Scouting and detection for pests arecrucial in quantifying the probable damage that may be caused by certain pests or implementing management approaches that are properly timed for maximum outcome. It also helps gardeners to determine when bugs become active or come to, the species present in a field, and their relative numbers as well.This procedure is also important for detecting preliminary symptoms of infection, which is crucial in flower transplants in order to avoid including infected flower seedlings into the garden, plus during the regular season in order to fine-tune management tactics as required such as controlling spray intervals. Usually, scouting includes monitoring for symptoms, like fungal colonies and leaf spots (Center for Agriculture, Food and the Environment, 2016).

7.7.2 Chemical Control

Chemical pesticides are typically used to manage diseases, weeds or pests. It's based on constituents that are toxic to pests involved. Once chemical pesticides get applied to protect the crop from vermin, they may also destroy the microbes living in the soil if properly applied.

Pesticides are frequently classified depending on the pest they're intended to manage. For instance, insecticides are applied for controlling insects; herbicides to crops; fungicides, fungi; and rodenticides, rodents. Additionally, some pesticides have chemosterilants or growth regulators that interfere with regular reproduction or growth (Center for Agriculture, Food and the Environment, 2016).

Chemical pest management has been applied for 1000s of years by many ancient civilizations, for instance, Sumerians discovered that sulfur provides great results in bug extermination.

7.7.3 Biology Control

Biological control approaches may also be applied as natural solutions since they don't rely on pesticides or other toxic chemicals. Natural biological pest management methods take a cue from the organism hierarchy within nature. It may be as modest as introducing a natural predator for the specific type of crop into your environment.This model of pest control isn't by any means modern. Societies have been using predators to manage pest populations from the 4th century B.C. Moreover, the fact it's still being used today shows exactly how effective this technique can be. The use of lady-bugs for controlling aphid infestations is a good example (Henderson, 2008).

7.7.4 Cultural Control

In floriculture, cultural control is referred to as a practice of changing the growing environment of a plant, in order to reduce the frequency of unwanted pests. Using cultural control methods, instead of chemical control, is beneficial in reducing or preventing harmful effects on the ecosystem immediately surrounding the garden. These approaches are aimed at decreasing the points of inoculum or reducing the plant exposure to contagion. Its primary goal is the prevention of vermin damage on flowering plants, but not the elimination of an already existing pest population. Case in point, good soil preparation and application of mulch means healthy plants that are fairly more resistant to vermin (Henderson, 2008).

8

MANAGEMENT PRACTICES FOR GROWING FLOWER CROPS

CONTENTS

8.1 Introduction 186
8.2 Environmental Factors For Growing Flower Crops 186
8.3 Soil And Growing Medium For Flower Crops 192
8.4 Optimum Spacing and Planting Density Of Flower Crops 195
8.5 Budding And Pruning Flower Crops 196
8.6 Optimum Nutrients For Flower Crops 199
8.7 Pinching Flower Crops 201
8.8 Disbudding Flower Crops 203
8.9 De-Shooting Flower Crops 205
8.10 Defoliation Flower Crops 206
8.11 Removal of Faded Flowers 208
8.12 Bending of Shoots Of The Flower Crops 210
8.13 Irrigating Flower Crops 211
8.14 Disease Control In Flower Crops 213
8.15 Insect And Pest Control In Flower Crops 216
8.16 Weeding 218

8.1 INTRODUCTION

Flowers have always been known to increase the aesthetic value of any place they are present. They turn bland areas into colorful eye-catching spaces. For flowers to achieve this, they need proper care and management. Some flowers are easy to take care of and others need attention. One needs to consider some things as they are choosing the flowers to plant.For starters, one needs to consider the type of flowers and whether the growing space available will be able to accommodate the proper growth of the flower. Whether an area is getting full sun or partial shade, whether the soil has good drainage and even whether the roots of your plant will have to struggle through clay soil.

One has to consider that perennial flowers have different maintenance conditions from annual flowers. Perennial flowers blossom every year for about 3-4 years until they finish their cycle and reach full maturity. Annual flowers mature in one year. Annuals require more aftercare than perennials. They require about ½ to 1 inch of water per week to be done in the morning so as to allow the water to dry out during the day. This prevents the growth of fungus. Perennials need mulch to conserve the soil moisture and maintain the soil temperatures.

Environmental factors such as light, temperatures, aeration, CO_2 and even the humidity needs to be considered. Deadheading as required is also important. This includes pruning, pinching, picking, sniping and cutting out dead flowers. This promotes the growth of more blooms by the plant. This write up will explain all these factors on how to manage growing flower crops.

8.2 ENVIRONMENTAL FACTORS FOR GROWING FLOWER CROPS.

8.2.1 Temperature

Temperature controls the timing of the crops. It dictates the time between the planting of the crops to the flowering of the plant. Plants grow flowers faster when grown in warm areas (an increase in temperature increases the plant development rate). Warm temperatures promote chlorophyll synthesis and development of the leaf surface area (Singh, 2006).

Seed development is also faster in warm areas. Plant growth regulators which are plant hormones that may alter seed characteristics increase in warm temperatures. One has to consider the flowers they are growing and the season. During summer, planting flowers that thrive in warm weather without wilting is important.

Figure 8.1: The effect of low temperatures on the leaves of potted plant. The leaves have been discolored and turned brown.

Source: https://www.gardeningknowhow.com/plant-problems/environmental/tips-for-saving-cold-damaged-plants.htm

Extremely high temperatures can cause stress on the plant. It can cause burning of the leaves and petals and may even cause the foliage to split. It may also cause metabolic disturbances by increasing the respiration rate, causing stunted growth. High temperatures can also bleach chlorophyll causing necrosis. Wilting leaves and seedlings are also a sign of extreme heat exposure, especially in the tropics. Therefore, one needs to know the correct temperature the flowers being grown need and this is the median temperature of 80 to 85 degrees F. Those that love heat cannot stand frost and any exposure to cold temperatures can kill the seeds of the crop. During planting, one should plant at a minimum soil temperature of 60 degrees F and avoid night temperatures that are lower than 55 degrees F (Singh, 2006).

8.2.2 Light

There are many processes in plants that are controlled by light such as germination, leaf expansion, chlorophyll synthesis, stem expansion, flower production and bud dormancy. Plants can be classified according to their light requirements (high, medium and low). Some flowering plants are affected by the number of hours they are exposed to light. There are short-day plants such as Poinsettias that only flower when exposed to light for 11 hours or less. Long day plants only flower when exposed to light for hours longer than 11 hours. There are others however that are not affected by the number of hours they are exposed. These are called day-neutral plants.

Increasing the amount of time that the plants are exposed to light also allows the plant to make enough food to thrive. Extreme exposure to light can also be harmful to plants. They should not be exposed to light for more than 16 hours in a day. Too much direct light causes the leaves to turn pale or even turn brown and die. The plants should be protected from excessive sunlight especially during the summer (Singh, 2006).

Figure 8.2: Effect ofextreme light causes which fading of flowers.

Source: https://www.gardeningknowhow.com/plant-problems/environmental/fixing-color-faded-flowers.htm

Fluorescent lights and incandescent light can be used as supplemental light. Fluorescent lights mostly produce blue light while incandescent lights mainly produce red and some infrared light. For plants to flower, they need all the light (red, blue and infrared). Despite the fact that plants require light to thrive, they also need periods in darkness to develop well.

8.2.3 Relative Humidity

Humidity levels are normally quite difficult to control. The levels fluctuate due to air temperature changes. Plants also constantly transpire thus increasing water vapor in the air. Humidity in the air causes the plants to constantly change the status of the stomatal openings of the leaves. High humidity causes slow usage of water by the plants which affect the quality of the crops. Low humidity causes the closing of the stomata to minimize transpiration thus reducing water loss. This however also reduces the rate of photosynthesis thus slowing plant growth.

When the humidity in the air is too low there are some negative effects:

- It causes crop wilting
- The crops have stunted growth
- They produce leaves with a small size
- The leave curl up
- It increases the probability of spider mites infestation

When the humidity in the air is too high there are also some negative effects:

- It causes soft growth of the crops
- It increases the chances of the crops getting a foliar disease
- Causes nutrient deficiencies
- It increases the chances of root disease and edema. It also makes the flowers susceptible to bud rot and botrytis

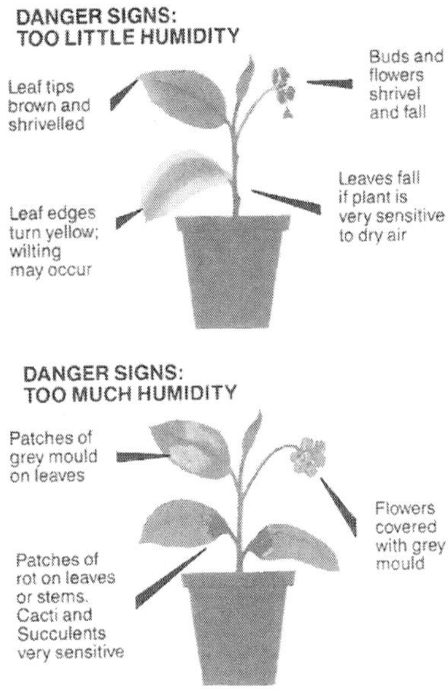

Figure 8.3: These are some of the effects of excessive or too little humidity.

Source: https://www.coolgarden.me/give-your-plants-the-humidity-they-need-3693/amp/

Flowering crops require 70-75% humidity. One can lower the humidity in the air by providing proper spacing between plants. This allows proper airflow and lowers the rate of condensation thus preventing the development of diseases like powdery mildew. One should also avoid overwatering the plants which cause damp soils and can increase the humidity. Improving ventilation also improves the airflow thus lowering the humidity levels (Singh, 2006).

8.2.4 Aeration

Poor airflow is the main factor that promotes dampness in the gardens. Good airflow promotes transpiration and evaporation of water. It also prevents the condensation of water on the leaves which can prevent rotting. It also allows the limbs to grow stronger by preventing softening from the dampness.

Damp gardens are a breeding place for many fungal diseases such as black spots on roses and powdery mildew. It also promotes the spread of diseases from one crop to another. When the soils are soggy, they act as a habitat for slugs and fungus gnats. Airflow can reduce and generally avoid all these problems.

There are a couple of ways to maintain proper airflow:

- One should ensure there is enough space from one plant to another to ensure proper airflow through the plants
- The plants should be exposed to the sun especially in the morning to evaporate the dew settled
- Avoid planting flowering crops in areas with high walls and hedges. Not only do they prevent light from reaching the crops but also block any breeze that is to reach the crops

Fans are recommended indoors for any indoor crops to improve the airflow. One can use a ceiling fan, an oscillating fan or even a floor fan. Indoor plants should also be kept close to a window which will give them both light and enough air flow. The plants indoors should also not be kept too close to each other to allow the available breeze to tousle the surfaces of the plant.

8.2.5 Carbon Dioxide

Plants take in CO_2 during the day and release CO_2 at night. Plants need CO_2 to ensure proper photosynthesis. The level of CO_2 in the air is measured in parts per million. One requires 390 ppm for proper plant growth. Any levels below 200 ppm hinder the growth of your plants. Increasing CO_2levels can greatly improve the growth of the plants. Having CO_2 levels that around 1000 to 2000 ppm can also allow you to grow your crops at high temperatures which will allow the plants to grow flowers faster (an increase in temperature increases the plant development rate). Adding CO_2 levels during the flowering stage especially in the first two to three weeks promote flower production and also boost the size of the flowers (Larson, 2012).

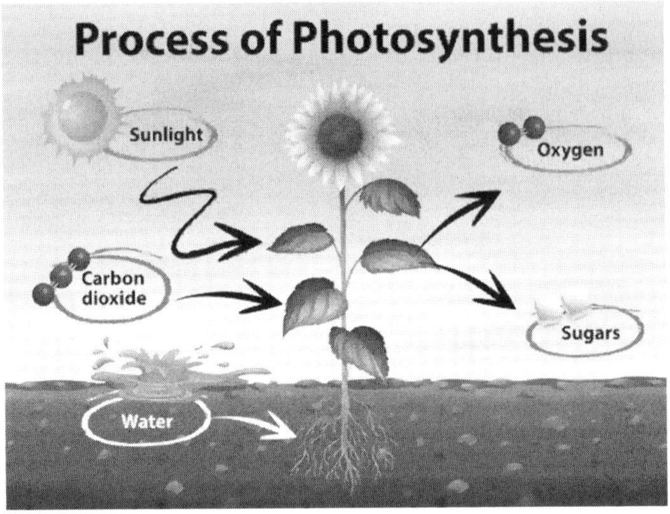

Figure 8.4: CO_2 is very important in the growth of flowers for without it there is no photosynthesis thus flowers cannot be produced by the plant.

Source: https://theconversation.com/amp/reclaiming-lost-calories-tweaking-photosynthesis-boosts-crop-yields-109283

There are a couple of ways to improve CO_2 levels:

- Burning propane or natural gas – these will release high levels of carbon. The only disadvantage is that it also increases the level of water and heat.
- Using a tank with a regulator to release CO2 at a certain rate – this can ,however, be less cost- effective because it has lesser carbon levels than propane and natural gas.
- Decomposition processes – Fungus and detritus bags can be used to provide CO_2 through decomposition. The material is costly but it produces large amounts of CO_2

8.3 SOIL AND GROWING MEDIUM FOR FLOWER CROPS

Soil has major functions for flowers' growth:

- It assists in the anchoring of the flowers to the ground
- It keeps moisture and provides water to the flower crops

- It is a source of nutrients to the flower crops

The soil has 4 components. It has water, air, organic matter and minerals. The soil has three functional phases. The liquid phase has hygroscopic water which is bound and not readily available to the plants, capillary water which is held by the capillaries and is available to plants and gravitational water which is held by tension and provides a lot of water to plants. The gas -phase contains air pockets with Nitrogen, oxygen and CO_2(Larson, 2012).

Proper management of the soil allows the soil to be at its peak performance improving the quality of the flowers. One should analyze the soil and check for the amount of humus in the soil, the value of pH and the amount of nutrients in the soil.

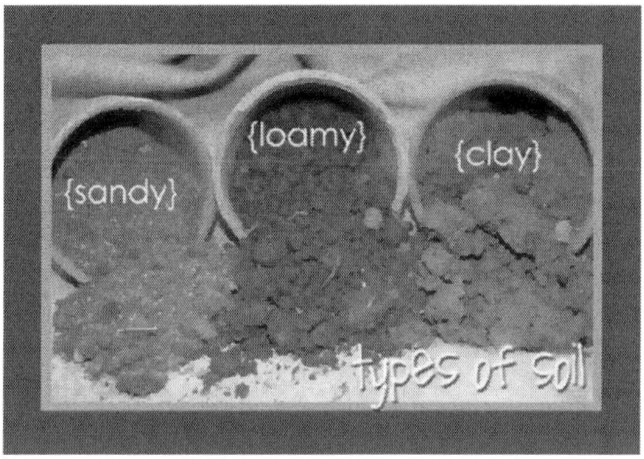

Figure 8.5: Different flower species prefer different soils to grow appropriately. The different soils have different properties.

Source: https://socratic.org/questions/which-soil-is-best-for-farming

The soils also require enough organic matter which will improve the soil structure and improve the ability of the soil to hold water and nutrients. It also prevents soil erosion and compaction assisting in promoting soil organisms. Good soil drainage is also important. If the soil does not drain water the soil becomes a dump and that is a good breeding place for fungi and other diseases.

Growing media is a material that plants grow in. It can be either solid or liquid. Its function is very similar to that of soil. The constituents of growing media are wood fibers, green waste, bark, peat, etc. Some mineral components including clay and perlite are used. Growing media is well blended and enriched with lime and fertilizers to achieve the appropriate chemical, physical and biological properties needed to grow flower crops.

For growing media to promote proper plant growth, the media requires the appropriate pH, density, porosity, aeration, organic matter and bulk density. Choosing the right media for floriculture is very important. It can affect the availability of nutrients and water, therefore, affecting the ease of growth of the flowering plants (Larson, 2012).

There are good options forgrowing media listed below:

- Soilless mixes – These are very common. They have a mixture of similar components: vermiculite and peat moss. Some include lime and even fertilizers. They come free of diseases and insects and have a good pH and drainage.
- Compost – These have a high content in macro and micronutrients. Despite the fact that they have good drainage and pH, the quality of the compost varies. One needs to be sure of the source. Compost should avoid being used as the main media but can be used as an addition to other media.
- Rockwool – This is made from melted rocks. It has good moisture retention and good structure however it has some disadvantages: it is not biodegradable, it tends to hold too much water which can cause root rot and the dust it produces is hazardous when breathed in. It needs maximum care to get the most out of it.

Figure 8.6: Rockwool growing medium. It is made from melted rocks and has very good water retention capabilities.

Source: https://www.advancednutrients.com/articles/using-rockwool-for-hydroponics/

8.4 OPTIMUM SPACING AND PLANTING DENSITY OF FLOWER CROPS

Proper spacing of flower crops is an important factor. It prevents competition for water and nutrients by the crops and also allows proper airflow. It also allows the crops to spread to their optimum width and reduces their vulnerability to diseases. The main way to know the exact space is to space the crops according to their expected width of maturity. This rule, however, can be broken if need be depending on the necessity of the crops. If flower plants are planted too close to each other, they become susceptible to diseases. The spaces provide airflow allowing the plants to dry out from one watering to another, preventing fungus which can cause root rot. This can weaken the plants, allowing insects to take advantage and feed on the crops (Larson, 2012).

Figure 8.7: Planting of perennial flower crop. The distance should be measured from the center of one plant to the center of the other.

Source: https://www.diynetwork.com/how-to/outdoors/gardening/when-and-how-to-plant-perennials

Perennials each have their own spacing needs, for instance, Hostas can grow 3 feet wide while coral bells grow 1 feet wide therefore one needs to measure the distance from the center of one pot to the center of the other pot. Annual plants, however, can have larger spaces between them because they have a longer time to grow giving them more to fill up the space.

Flower crops that are planted close to each other require more fertilizer to provide for all the roots clustered in that area. Ground covers, for instance, are planted quite close to each other so as to cover the soil faster. They spread continuously so the spacing does not really matter (Larson, 2012).

The plant density of a flower bed is described by the number of plants within a given unit of area. Flower beds with low density make them more susceptible to weeds and those with high density may lead to diseases such as root rot.

There are some reasons plant spacing is important discussed below:

- Nutrient intake – Flower crops that are grown too close to each other compete with each other for both nutrients and sunlight. This denies them the opportunity to be as healthy as those that are not competing with any other crops.
- Disease management – Maintaining Spacing of the flower crops prevent transmission of diseases from one crop to another and it also improves their immune system. Planting the crops too close to each other affects the airflow. This can also make crops susceptible to diseases.
- Weeding – Planting crops too close to each makes it harder to pick out and uproot weeds, thus resulting in more intensity of weeds.
- Increases the number of flowers – Proper spacing allows for flower crops to grow to their maximum potential by allowing space for the flowers to grow.

Good spacing can be important for keeping the soil moist and can also crowd out the weeds. But this only works if the crops are spaced appropriately. When you buy a plant, it comes with certain spacing guidelines that should be followed. Proper plant spacing is important in improving the quality of the yields thus getting value for money. It is also advisable to plant them in rows. This makes it easier to harvest the crops.

8.5 BUDDING AND PRUNING FLOWER CROPS

Budding is the process of taking a bud plant and growing it on another plant. It is a vegetative method of propagation that leads to the growth of a similar plant from which the bud was taken from. It can only be performed when the weather conditions and the physiological stage of the flowering plant are both optimum. Budding requires a lot of care and experience. One needs to make sure that the scion that is collected has buds that are mature, that the rootstock matches with the cambium of the scion and that the rootstock can be compatible with the scion. One needs to make sure not to contaminate the bud.

Figure 8.8: Buddinng in Rose - Budding is the process of taking a bud plant and growing it on another plant. The bud grows to be similar to the plant from which it came from.

Source: https://www.youtube.com/watch?v=lmViA-kMwvM

Budding may increase the productivity of flowering crop by improving certain qualities:

- Change in variety – New varieties may improve disease and insect resistance. They may also produce a higher number of flowers and tolerate drought better.
- Provide optimum cross-pollination and pollination – Some flowers do not self-pollinate and require pollination from different flowers of a different variety. Some flower plants are also dioecious thus have only one reproductive part, male or female. The chances of cross-pollination can occur if a bud is taken from the male plant and grown on the female plant.
- Repairing of damaged flowering plants – Repairing the damage can be done by planting several seedlings of a similar species to the one that is injured and growing them above the injury.

Pruning is the method of selectively eliminating a specific part of a plant so as to improve the health of the whole plant. Pruning can be done to remove any injured, damaged or sick plant parts, to improve the quality of the flowers, to improve airflow (which can prevent diseases), to promote plant health and to train the flowering plant. Plants that flower during the spring should be pruned after flowering. They normally produce their buds the year before during the winter and they open up during spring. Pruning before they flower (during fall or winter) will lead to removal of the buds thus there will be no blooming of flowers during the spring (Henderson, 2008).

Plants that flower during the summer should be pruned during late winter or the early spring. This promotes the growth of new and more flowers in the coming summer. Newly planted flowering plants should not be pruned unless they have been damaged because they require as many leaves as possible to promote growth of the plants, especially the roots.

Figure 8.9: Gardener cutting off the spent flower so as to improve the aesthetic value of the blooms. The spent flower has withered petals and that makes the cluster inconsistent.

Source: https://www.simplemost.com/easy-guide-pruning-plants-beautiful-flowers/amp/

There are two different methods of pruning: thinning cuts and heading cuts. Thinning cuts remove the entire limb to the point from which it starts while heading cuts remove only a part back to the bud.

Flowering plants like the Columbine are highly self-seeding which can be a problem in a contained area. They produce many seeds that multiply on their own. Columbine flowers should be pruned in the fall to avoid and prevent self-seeding (Henderson, 2008).

8.6 OPTIMUM NUTRIENTS FOR FLOWER CROPS

The provision of nutrients to flower crops is important and it promotes a good yield of high-quality flowers for sales. This should be achieved without contaminating the environment and maintaining the soil's health by using fertilizers in an appropriate manner.

Some of the major nutrients are:

- Nitrogen – this is one of the hardest nutrients to maintain in the soil. It is required in large amounts by plants but it is also very easy to lose in the soil from leaching or as a gas. Low nitrogen levels in the soil result in lower flower yield, poor plant growth and the formation of foliar chlorosis. Excessive nitrogen can, however, cause the formation of weak stems, delayed flowering and excessive vegetative growth. Nitrogen fertilizers should be applied at least once a year to ensure proper production. Nitrogen leaches from the soil as nitrates which can pollute water bodies. Nitrogen also influences soil pH. Ammonium fertilizers can cause soil acidity when the ammonium is converted to NO_3 by bacteria in the soil. Lime can be used to neutralize the acidity.
- Phosphorus – low levels of phosphorus may cause the plant to be small in stature and also produce short stems. It also promotes the flowering of plants. Phosphorus, however, causes negative effects onthe environment by running off to water bodies causing the growth of algae which lowers the water quality.
- Potassium – low potassium levels can result in marginal chlorosis and the burning of the leaves on the lower part of the plant. Excessive deficiency may affect the leaves on the flowering stems and reduce the yield. Low potassium levels are mainly seen in soils with high sand levels and those with little to no organic matter. Soils having little to no organic matter have low cation

exchange capacity and can therefore not retain positively charged ions such as potassium, magnesium and calcium.

Figure 8.10: Eggshells can be used to provide nutrients to the soil as a form of compost. They decompose and release nutrients.

Source: https://www.mnn.com/your-home/organic-farming-gardening/stories/amp/how-make-homemade-plant-food

During fall or spring, one should test the soil for its levels of nutrients and pH. If necessary, limestone can be added to the soil. It is important to maintain soil pH at 5.5 to 7 so as to maintain soil fertility. The pH affects the presence of trace elements such as phosphorus which is highly available at a pH between 6 and 7. Limestone can also act as a fertilizer for magnesium and calcium (calcitic limestone with 5-10% magnesium). Liming also prevents the buildup of aluminum and manganese which can be toxic.

Low nutrient levels in the soil can be noticed on a plant with following unusual characteristics:

- Early stunting of the crop.
- A noted difference in growth compared to a crop growing close by.
- Premature or late maturation.
- Slow growth.
- Yellowing of leaves.

- Poor quality flower production.

Earthworms can also help improve nutrient levels in the soil. They feed on materials in the soil that are either dead or decaying and thereafter excrete dung that is rich in nutrients. Providing good compost to your soil can help attract earthworms. This is the art of vermiculture and vermicomposting. Liming can also promote earthworm activity and the rotting of organic matter in the soil, releasing N, phosphorus and micro-elements (Henderson, 2008).

Figure 8.11: Earthworms are very beneficial in improving soil nutrition. They feed on materials in the soil that are either dead or decaying and thereafter excrete dung that is rich in nutrients.

Source: https://www.farmersweekly.co.za/crops/field-crops/benefits-of-vermiculture/amp/

8.7 PINCHING FLOWER CROPS

Pinching flowers refers to the cutting off the top part of a flower stem. When you remove just the top part of the growing tip, the apical meristem, that is called a soft pinch. When you remove a couple of inches from the top of the stem together with some leaves, that is called a hard pinch. These two actions can prolong bloom and increase yield (Henderson, 2008).

Once you cut off the tip of your crop that is growing without giving an opportunity to bloom, the plant is stimulated to grow multiple stems that grow flowers at the same time. Pinching can help lengthen the time in which certain flowers bloom. Pinching should be done as close to the nodes of the lead as possible with your fingers or nails.

Deadheading refers to removing spent flowers. This allows flower plants that repeat-bloom to continue blooming. If the spent flowers are not removed, the plants produce seeds instead of flowers. One should not pinch off old flowers if they wish to use the seed to plant some more flowers or provide seeds for birds. Perennials that self-sow should have the flowers pinched off regularly to prevent the growth of excessive seedlings. Blooms or stems can be pinched off once or a couple of times during the growing season. In scabiosa plants, it is advisable to deadhead the whole flowering stem because each flower grows on itself own stem.

Figure 8.12: Pinching is very beneficial to flower plants. This is an example of a spent flower being pinched out at the stem. This provides bigger blooms.

Source: https://www.prescottenews.com/index.php/features/columnists/mountain-gardener/item/27688-pinching-for-bigger-blooms

Some plants known for their foliage such as the coleus plant can benefit from pinching off the flowers. It encourages bushier and fuller plants. Perennial flowers that bloom in the fall are pinched earlier in the season to prevent excessively tall plants and promote the growth of more flower buds. Mums can be pinched to delay their bloom, allowing them to bloom when other plants are dying down. Sometimes the climate at hand can compensate for spent flowers that are not pinched. Wind or cold temperatures can kill the growing tip causing the plant to branch lower on the stem and grow back looking fuller than it was before (Henderson, 2008).

Advantages of pinching include:

- It grows a fuller plant.
- There is the production of more desirable flowers and leaves.
- It slows or lengthens the blooming period.
- It works as a good alternative to cutting back.
- It helps eliminate stressed flowers so as to divert that energy into producing more flowers in flowering stems that are capable.

Pinching can be used to train a flowering plant. Training helps achieve some positive effects:

- It protects the flowers from wind damage.
- To expose maximum leaf surface to the sunlight for optimum photosynthesis.
- To ensure there is a balanced distribution of flowers on the flowering stem.

Some of the flowering plants that benefit from pinching are petunias, sweet peas, geraniums, marigolds, fuchsias and dahlias. Every time a plant is pinched it lengthens the blooming time resulting in more side branches but smaller flowers. Most cooking herbs benefit a great deal from pinching. Regular pinching of rosemary can help keep the plant at a manageable size, especially indoors(Dahal, 2013).

8.8 DISBUDDING FLOWER CROPS

This process is used to limit the number of buds growing by gradually removing them as they grow on the flowering stem so as improve the quality of the remaining buds and also improve their lifespan. It is a very beneficial process for weaklings and younger plants. One should choose the bud to remove by noting the diameter of the bud and making sure it is equal to the size of a dime or smaller. It is not advisable to remove any buds larger than that. When a cluster of blooms undergoes disbudding, it produces larger blooms in the cluster, creating a very full bouquet of flowers. This also occurs if the main center bud is removed from the cluster, making the remaining buds open at around the same time.

I will use a rose bush as an example. When you buy a bush and transplant in a garden it goes through a period where its roots are trying to establish a working system in the new environment and also supply nutrients to the growing blooms. Doing both at the same time may take a lot of the plant's energy thus putting stress on the plant. Removing all the buds and blooms on the plant at the time allows the plant to start by establishing its root system then produce beautiful and healthy blooms thereafter.

Disbudding roses can improve the quality and size of the bloom. Using the thumb, one should pinch the small buds that develop below the main bloom from that angle between the leaf and the stem and move them front and back until they fall off. This should be done earlier in the development of the bud to avoid leaving scars and black stubs (Singh, 2006).

Figure 8.13: Disbudding helps in providing a larger bloom and making the bouquet look fuller. The central bud should be left intact as seen above.

Source: https://www.floristwithflowers.com.au/blog/disbudding-for-the-perfect-bloom/

Disbranching is a form of disbudding that is mainly used in Dahlias plants which produced numerous shoots that have to be managed to prevent overcrowding. It is also beneficial to control the number of flowers per stem so as to secure good blooms from large flower varieties. This can be done by removing the small buds around the central one and leaving the central one intact. It is recommended that this process be done slowly by slowly spreading it out in a week compared to removing all the buds at once.

Camellias tend to drop buds as they grow. This can be stopped by disbudding especially when you notice that the plant has an unusual number of buds. Lilacs can benefit from disbranching when the young weak shoots are removed. This directs all the energy to the main branches that improving the quality of the bloom.

Disbudding achieves a couple of things listed below:

- It helps direct energy to the development of fewer flowers thus improving the quality and size of the bloom.

- It helps dictate the position and space between blooms to allow the flowers to blossom without any hindrances.
- It protects the blooms from danger by reducing the chances of damage from moving twigs and leaves that may have been close to them.

8.9 DE-SHOOTING FLOWER CROPS

This is the removing of the terminal bud of the flowering shoot and continuous removal of the lateral buds from the lower axis of the leaf. This process can improve and increase flower production and improve structural shoot formation. Side shoots can easily be spotted. They are located between the leaf and the stem. It is also considerably thinner than the stem. It can be pinched with the finger or cut with scissors. This has to be done as close to the stem as possible.

In the case of other crops like tomatoes, there is a possibility of having a twin head. In this case, the weaker one should be de-shooted. There are also times when the plant has no head leaving a plain truss. A shoot on the lower side can take up the role of being the main stem. The shoot without the head can be cut off.

Figure 8.14: Removing shoots can help increase the number of flowers that grow per shoot, giving a fuller bouquet.

Source: https://www.wikihow.com/Prune-a-Gardenia-Bush

De-shooting allows for the plant to direct its energy to producefuller flowers with better quality. It can also be done if the shoot has been infested by diseases or pests, causing the plant produce poor quality flowers. Aphids can suck plant juices from the shoot, making the shoot weak and stunt. Stunted shoots should be cut off so as to allow the plant to focus on its own health (Singh, 2006). De-shooting stunted plants also promote growth of healthier shoots that will produce healthy and larger flowers. It also helps produce more flowers per shoot.

8.10 DEFOLIATION FLOWER CROPS

Removal of leaves from the flowering plant, it. can be done by nature or manually.

The weeping fig is a good example of nature's way to defoliate. It drops more leaves as it grows. This drop happens because the plant is not receiving enough sunlight to support the number of leaves on the plant. This is the plant naturally repairing itself allowing the plant's energy to focus on improving the plant's health.

Manual defoliation can be done to minimize damage or disease. Environmental factors could cause the yellowing, fading, damage or spotting of leaves. These can also be symptoms of plant diseases or pests. Such leaves make the foliage very unattractive and need to be removed at some point (Ravinath, 2007).

Figure 8.15: Leaves that are discolored and browned should be defoliated so as to make the foliage more attractive.

Source: http://www.27east.com/news/article.cfm/East-End/478989/Tips-To-Make-Roses-Thrive

If more than half of a leaf is yellow, damaged or dead, it should be removed by grabbing the base of the leaf where it is attached to the stem and breaking it off. The entire leaf should be removed without leaving any part, so as to avoid areas that pests would prefer to breed in.

Some plants have hard parts that require scissors to remove. For large leaves such as that of the Warnecke, one can tear the leaf right at the center where the midvein is, all the way to the trunk. The leaves can then be easily removed from the trunk with a tug. Palm leaves are quite large and require scissors to remove. Their size might be a reason to see browning on the tips of their flowers. This should not raise alarm. The brown parts can be trimmed off while naturally shaping the leaf. Avoid trimming the whole brown section as leaving a little will prevent further browning of the leaf.

Removing leaves is not only important when the leaves are damaged or ill, but also to maintain the health of the plant. Hanging plants may require defoliation so as to allow them to be well balanced and shaped. Hanging baskets can also be defoliated to keep them short so as to avoid affecting any living or working area beneath them. Cutting leaves off can also promote the growth of the plant in other areas when energy is low and needed to make flowers (Ravinath, 2007).

Figure 8.16: This is the Schwazzing technique which involves removing fan leaves to allow light to reach the buds below.

Source: https://www.advancednutrients.com/articles/removing-fan-leaves-schwazzing/

Some plants such as daffodils should not be defoliated until six weeks after their flowers have grown. Since it is a bulb plant, it requires to undergo photosynthesis via the leaves so as to provide enough energy for the bulb to store to produce blooms the following year. After that, the leaves can be removed before they start to yellow and die. Since they are outdoors, they should be cut off before they fall on the ground and cause a pile of soggy leaves, especially during the wet season. This can be a breeding site for fungus and other diseases. The leaves should only be allowed to fall if the plants are planted under trees with a large shade to prevent them from getting wet. In this situation the leaves can rot down and act as compost for the plants.

8.11 REMOVAL OF FADED FLOWERS

This process of removing faded flowers is referred to as deadheading. It should be done all through the growing season because many flowers lose their attractiveness by spoiling or fading with time. When the flowers start shedding their petals, they create seed heads thatconsume all the energy in the plant. This leads to lack of flower formation. Deadheading can be used to channel the energy to growing flowers instead. This also improves the quality of the flowers.

Some of the flowers that benefit from deadheading are phlox, bleeding hearts, sage, coneflower, salvia, etc. There are some advantages to deadheading listed below:

- It makes the plant look neater – when the flowers are spent, they turn brown and dry up or become very soggy. This can make the garden look unkempt.
- It encourages the plant to grow more flower buds – plants mainly flower to produce seed heads. Deadheading regularly before they mature just leads to the formation of more flower buds so as to set seed. This keeps the flowers blooming. This works well for flowers like petunias and marigolds.
- Helps the plant conserve energy – removing spent flowers allows the energy in the plant to focus on creating healthier blooms and improving the general health of the plant. Peonies bloom once a year even when the spent flowers are removed but when you cut back the stalks of the flower, the energy in the plant goes to improving the root system and the plant foliage thus improving the health of the plant by allowing the plant to regain its energy that it may have lost during the blooming period.

- It prevents seed formation – there are plants that self-sow, and this can cause overcrowding. Deadheading prevents the seeds from setting.

Deadheading should occur from spring to fall especially when the flowers start to fade. Cutting close to the bloom leaves an unattractive stem. Deadheading should be done about a quarter-inch above a new lateral flower, leaf or bud encouraging healthy foliage. The spent flowers you cut off can be collected and added to compost. Despite the fact that deadheading promotes new growth, one still needs to fertilize the soil regularly to ensure the plants grow strong and healthy especially for annuals which are heavy feeders.

Figure 8.17: Faded flowers should be removed so as to allow the plant to use its energy on improving the health of the other flowers and the plant itself.

Source: https://www.gardeningknowhow.com/ornamental/flowers/fgen/pulling-dead-and-faded-flowers-off-plants.htm

Flowers provide nectar to bees, butterflies and birds which is more than just beautification. Deadheading provides more flowers and healthier flowers with more nectar and pollen for pollinators to indulge in. Deadheading is important but we have to consider as well that the plants produce flowers to produce seeds that are important in creating new generations of the flower species. So the gardener must follow his/her gut and know when and when not to deadhead (Ravinath, 2007).

8.12 BENDING OF SHOOTS OF THE FLOWER CROPS

When the flower crops are planted, they should not be allowed to flower immediately because it can affect the structure of the plant. To avoid this, the first flower is pinched off a month after planting. This will cause two to three eye buds to sprout on the main branch. These will develop to become branches. These branches will later develop to form buds. At this point is when the shoot is bent near to the bud joint toward the path the plant is taking. This is done to initiate bottom break a ground shoot. The surface area of the leaf isimportant in building a strong root system. The surface area needs to be at its maximum (Ravinath, 2007).

Development plays a key role in the plant structure and in developing multiple growing points. After planting, a ground shoot begins to grow at the crown of the plant. This ground shoot is weak and should be bent at ground level. This helps the plant in the formation of a strong plant structure so that it can produce quality flowers all through its life cycle.

Bending helps attain and maintain the maximum leaf area. This area helps to build astrong root system that is required if the plant is to remain healthy. The leaves are the primary source of carbohydrate production.

$$6H_2O + 6CO_2 \longrightarrow C_6H_{12}O_6 + 6O_2$$

This is the formula for photosynthesis. The "lungs of the plant" are actually compared to the mass of leaves. To check for new growth on the bent stem, one must remove the suckers that are growing. One should also not forget to check for bud rot or thrips on the bent shoot by removing the buds on the bent stem. One should only select the weak shoots for bending. This process should be carried out throughout the life cycle as it is a continuous process. It also breaks the apical dominance of the plant. It should be done in a manner that leaves most of the stems laying below horizontal (Henderson, 2008).

This process is not advisable in the summer as it promotes an incidence of mites. It is good to note that one should bend on the first or second five pairs of leaves only. Care should be taken while bending the stems to ensure that one does not break the stems and the leaves do not touch the soil. Roses can be grown in a greenhouse without bending the shoot. One can keep some blind shoots standing upright so as to increase the rate of photosynthesis and the rate of water nutrient uptake by the plants.

Figure 8.18: Bending helps increase the leaf area, allowing the plant to build a strong root system for acquiring nutrients for the plant.

Source: https://www.researchgate.net/publication/40095115_Crop_theory_and_model_in_the_era_of_systems_biology

While harvesting, there are some stems that should not be harvested: disease and pest infected stems, bent stems and stems that are too weak or too thick. These stems should also be bent so as to increase the leaf area of the plant (Henderson, 2008).

8.13 IRRIGATING FLOWER CROPS

Watering flowers is important but it has to be done very well and with the utmost care. Excessive watering can cause the growth of fungi and cause root rot which can kill the plant. For proper flowering one must consider the weather and the type of soil.

Most flower gardens only require one inch of water every 7-10 days. This can be achieved in areas with rain or using irrigation methods. Flowers that are grown in dry hot areas, however, require more water. Deep watering is preferable because it allows the roots to grow long, strong and healthy compared to shallow watering that keeps the roots shallow and it promotes water loss through evaporation. Flowers that are grown in sandy soil require more watering because the soil having loose particles does not retain water. The soil should be allowed to dry up in between waterings to prevent root rot (Henderson, 2008).

Timing is also very important. It is advisable to water flowers in the morning or the early afternoon to allow the flowers to avoid drying up during the hot afternoon sun. Watering in the late afternoon will evaporate the water, preventing the flowers from absorbing any water and watering at night can make the plants retain too much water and that can be a breeding site for fungi and other diseases. Lawn sprinklers are easy to use but they are quite inefficient. Approximately half of the water or more can be lost through evaporation. On top of that, wetting the flowers and leaves makes them soggy and this can make them susceptible to diseases and pests. Watering by hand can also be very inefficient. It still only waters the top of the soil which is shallow, promoting evaporation but also leads to the formation of puddles on some areas due to lack of consistency and even drier parts in other areas. Puddles can be a habitat for slugs that can breed there and even feed on the plants.

The most efficient way is by using a soaker hose. This allows the water to slowly seep into the soil without wetting the flowers and leaves and allowing the roots to reach deep into the soil for the water, making them deeper.

Watering potted flowers is different. The soil in containers dries up faster than soil that is found in the garden. Containers that are kept outside should be checked for dry soil daily, especially during the summer. They should only be watered if the soil is dry to touch. If the soil is still damp from the previous irrigation, they should be left to dry up before the next watering to avoid making the soil soggy which can be harmful to the plant's health. If the top one inch is dry to touch, the plant should be watered until the water is pouring out through the draining holes at the bottom of the pot or container (Henderson, 2008).

Figure 8.19: Sprinklers are not advisable in irrigating flowers because they wet the foliage, making the plant susceptible to diseases and pests.

Source: https://gardenerspath.com/gear/irrigation/methods-home-gardens/

In large gardens, drip irrigation is an efficient method because it allows the water to seep into the soil. It is done by installing plastic pipes in the soil which deliver water directly to the roots. This method helps conserve water as it does not use much water because the water doesn't get a chance to evaporate as it is soaked in almost immediately. The fact that it does not wet the entire plant also makes it appropriate. Wetting the whole plant will lead to fungal infections.

8.14 DISEASE CONTROL IN FLOWER CROPS

There are three extremely common flower diseases: powdery mildew, black spot and grey mold which is also called Botrytis. Here are ways to notice, control and prevent these diseases.

Powdery mildew: This is a whitish powdery layer found on the surface of leaves. In extreme cases, it might also appear on the stems and the flowers themselves. Once a leaf is affected, it discolors and turns yellow then brown and falls off. This disease is not fatal but it is very unattractive and if the disease occurs multiple times, it weakens the plant. Not only can it be spread by water, but also high humidity.

Powdery mildew can be prevented in a couple of ways given below:

- Proper spacing and eliminating weeds can help improve airflow and maintain appropriate levels of humidity.
- Overhead watering should be avoided to avoid increasing the

humidity and to also avoid wetting the flowers and leaves. Watering should be done at the soil level.

Figure 8.20: Fungus can spread on the whole leaf appearing as a white powdery layer symptoms of powdery mildew.

Source: https://www.greenlife.co.ke/powdery-mildew-of-cucurbits/

On top of preventing powdery mildew, we can also control it:

- Choosing flower cultivars that have been proven resistant to powdery mildew.
- Once you notice any of the leaves with the symptoms, they should be plucked off and thrown away.
- Insecticidal soap can also help if applied immediately the symptoms occur and should be repeated during the growing season.

Blackspot: This is a common fungal infection mainly seen on roses. It causes black spots on the foliage especially the leaves on their upper surface. A yellowing then follows around the spots and the leaves become fully yellow and drop off the plant.

This fungal disease can be prevented and controlled in a couple of ways as given below:

- Keeping the foliage dry is a primary way to prevent blackspot. Despite the fact that it takes about 7hrs of the foliage being wet to cause disease, it is better to keep the foliage dry. The fungus is also inhibited at temperatures of about 85 ^{0}F.

- Any susceptible flowers such as roses should be planted in a location with enough sun to keep the foliage dry. They also require a nice open space with good airflow.
- Any infected leaves should be pruned out and disposed of.
- Choosing resistant rose varieties.

Gray Mold: This is an extremely problematic fungus during rainy seasons. It appears as water-soaked spots and progresses to a gray fuzzy coating on spent leaves and flowers.

This can be prevented and controlled in a couple of ways as follows:

- Any infected leaf or flower should be plucked out and disposed of.
- Some beneficial bacteria can prevent the growth of gray mold.
- Eradication of weeds can help improve airflow which is paramount in preventing gray mold.
- Watering at soil level prevents wetting the foliage which will help keep this fungus aways.

Figure 8.21: Wetting the flower plants' foliage can lead to the growth of this fungus. It can be very problematic during the rainy season.

Source: https://pnwhandbooks.org/plantdisease/host-disease/chrysanthemum-gray-mold

8.15 INSECT AND PEST CONTROL IN FLOWER CROPS

Insects and pests feed on the plant leaves and cause yellowing, destroyed foliage and damaged blooms. These insects and pests can be avoided by pesticides but this can also affect beneficial insects. There are alternative measures that can be used to control insects and pests.

Aphids: These are pests that are shaped like a pear. They suck sap distorting foliage and causing yellowing and falling off of leaves. Feeding on the sap allows them to spread viral diseases to the plant. They also excrete honeydew on the leaves which can make them susceptible to mold. They can be controlled by as listed below:

- Strongly spraying the plants with water.
- Natural control by encouraging other insects such as lady beetles and aphid midges which will feed on the aphids.
- Spraying repellents with garlic or hot pepper keep aphids away.
- Insecticidal soap can be used in extreme cases.

Figure 8.22: Aphids attack rose flowers leaving it looking very unattractive and it can kill the flower crop.

Source: http://anneofgreengardens.com/blog/aphids-vs-ladybugs/

Caterpillars: These chew on the flower leaves or along the margins. To eliminate them, the following can be followed:

- One can use floating row covers to keep them away.
- One can also encourage other predictors that will eat the caterpillars.

Cutworms: These are wide segmented larvae that can be either black or gray. They are mostly active at night. They feed on the plant's stem and can completely eat through small plants. They can be controlled by:

- Since they are mainly observed in May and June, one can delay planting.
- One can handpick them from below the soil surface.
- Using cutworm collars is important fortransplants.

Japanese Beetles: They come in a metallic bluish greenish color with wings that are bronze in color. The adults chew on leaves and flowers and can end up defoliating the plant. The larvae which are wide and white with brown heads feed on the plant roots. They can be controlled by:

- Shaking them off the plant every early morning.
- Using insecticidal soap on the plants.
- Floating row covers may also help.

Scales: These suck the sap from the plants thus weakening them. It causes yellowing of the leaves which causes them to drop and die. They also excrete honeydew which can make them susceptible to mold. These can be controlled by:

- Removing any infected leaves by pruning them out.
- Spraying neem oil on the leaves can help keep them away.
- Predators that can feed on the scales can be encouraged.

Figure 8.23: Japanese beetles feed on flowers leaving them looking like unattractive.

Source: https://gardenerspath.com/how-to/disease-and-pests/japanese-beetles/

Tarnished plant bug: These bugs move quick and come in either green or brown. They have wings with yellow triangles. They suck on the plant juices and cause stunting and wilting of the plants. They can be controlled by:

- Eliminating weeds and promoting airflow.
- Encouraging predators that will feed on the bugs.
- Floating row covers can also keep the bugs away.

In extreme cases of infestation by these insects and pests, one can use insecticides and pesticides as the last resort. Pesticides can be harmful to our bodies through ingesting, inhaling and even absorption. One needs to be careful and wear personal protective equipment.

8.16 WEEDING

Weeding is a tiresome and time-consuming job but it is very beneficial to the flower plants. Weeds compete with flower crops for nutrients water and even light thus having lower flower yields and making the plants susceptible to diseases and pests.

Figure 8.24: Weeds compete with flower crops for nutrients water and even light thus having lower flower yields and making the plants susceptible to diseases and pests.

Source: https://gilmour.com/how-to-prevent-weeds-garden

Different weeds emerge in different seasons. Those that grow during the summer and those that grow in the winter. Herbicides are available but are not always efficient in controlling weeds. Cultural practices can be used to effectively control weeds. Some of the cultural practices are as explained below:

Field preparation: Preparing the field before planting can help keep weeds at a minimum. Tilling the soil two to three times early before planting will bring the seeds of the weeds to the surface where they will germinate and be tilled under in the next tilling process. This should be done in the middle of July, and afterward, the weeds should be allowed to grow from July to fall. Thereafter, glyphosate can be applied to kill the weeds finally making the soil ready to plant in.

Cultivation and spacing: Planning the plants early before weeds grow can give the crops an advantage inutilizing the nutrients thus avoiding competition with the weeds. If the growth of the crop is affected in any way, the weeds can take over and utilize ask of the soil's nutrients.

Mulches: Organic mulches effectively control weeds. Compost such as rotted sawdust, hay or straw hasproven to work. They, however, need to be composted correctly because if not, they can reduce Nitrogen levels in the soil. Hay, however, may have too many weed seeds and increase the chances of weeds growing. Mulching should be done to warm soil without any weeds immediately after planting the crops.

Chemical control: Herbicides are classified into two: pre-emergence and post-emergence. Pre-emergence herbicides are applied after planting the crops but before any weed seeds have germinated. Post-emergence herbicides are applied after the weed seeds have germinated and emerged. Selectivity is also another way to classify herbicides. Nonselective herbicides are not picky and control any weeds that emerge. Selective herbicides can only control specific weeds. Glyphosate is an example of a nonselective post-emergence herbicide. It works well on actively growing weeds. It is good for field preparation as well. Sethoxydim is a selective post-emergence herbicide. It controls many annual and perennial weeds. When applying it to cut flower crops, any blooms should be harvested to avoid injury to the blooms. Bensulide is a pre-emergence herbicide. The flowers need to be fully mature before the application of this herbicide. Some of the flowers that are safe when this herbicide is used are daffodils, daisies, tulips, wallflowers, sweet peas, gazanias, primrose, etc.

To use any of these methods, one must take into account the cost of labor and how available and affordable the materials are. Surveying the site and noting the weed pressure is also important so as you can know what method suits the site perfectly.

9

HARVEST AND POST-HARVEST HANDLING OF FLOWER CROPS

CONTENTS

9.1 Introduction..222
9.2 Pre-Harvest Factors Affecting Flower Longevity. 223
9.3 Harvest Factors Affecting Flower Longevity..225
9.4 Post-Harvest Factors Affecting Flower Vase Life..................................228
9.5 Harvesting Flowers...239
9.6 Flower Storage Methods...240
9.7 Storage Conditions For Flower Crops..244
9.8 Grading And Sorting Flowers..247
9.9 Bunching..248
9.10 Chemical Solutions..249
9.11 Packing..251
9.12 Cooling..253
9.13 Irradiation..258

9.1 INTRODUCTION

For high-quality yields, one has to follow the guidelines provided for harvesting flower crops. It is paramount in ensuring thelongevity of the flower crop. It is influenced by pre-harvest procedures, weather conditions and plant environment.

Before harvesting, a few boxes have to be ticked:

- The plants being harvested need to have reached their maximum turgidity and should be healthy.
- Any equipment including the buckets or containers and the cutting tools should be clean and sanitized. The water used inside the buckets needs to be clean.
- All the cutting tools should be well sharpened to avoid crushing the stems which can hider proper water uptake.

Figure 9.1: Harvesting is best done in small quantities at a time and the gardener should ensure that the flowers are immediately taken to a cool area to maintain their quality.

Source: http://slowflowersjournal.com/index.php/2018/08/26/bloomwithjohnnys-cutting-garden-inspiration-and-resources-from-johnnys-selected-seeds/

Once all these boxed are ticked, harvesting should be done in the morning or in the evening. While harvesting, any foliage that is below the water should be removed and the cuts should be slanted which helps increase water intake. Harvesting is best done in small quantities at a time and the gardener should ensure that the flowers are immediately taken to a cool area to maintain their quality. Below are some harvest and post-harvest methods of properly handling flower crops.

9.2 PRE-HARVEST FACTORS AFFECTING FLOWER LONGEVITY.

9.2.1 Genetic or Inherent Make Up

Genetically modified flowers are produced from other flowers that have had their DNA introduced to certain changes. These allow the flowers to exhibit new traits. It also gives more control over how one wants the flower to be. Genetic diversity has caused differences in the vase life of cut flowers. Rose cultivars such as black tea and Leonidas have a vase life of 4 to 5.8 days whereas cultivars like Gold strike and Danya have a vase life of 12 to 14.5 days (Center for Agriculture, Food and the Environment, 2016).

The vase life of a flower correlates with the concentration of sugar in the petals. Some varieties are not able to open once harvested at bud stage (Grand gala) due to low carbohydrate content, however, cultivars such as bridal oink have not shown any problems with bud opening when harvested before the buds open.

9.2.2 Environmental Factors

Factors such as temperature, light, humidity, Carbon dioxide, season, pests and stage of maturity are primary in the increasing post-harvest longevity.

- Temperature - Extremely high temperatures can cause stress on aplant. It can cause burning of the leaves and petals and may even cause the foliage to split. It may also cause metabolic disturbances by increasing the respiration rate, causing stunted growth. High temperatures can also bleach chlorophyll causing necrosis. Wilting leaves and seedlings are also a sign of extreme heat exposure, especially in the tropics.
- Light - There are many processes in plants that are controlled by light such as germination, leaf expansion, chlorophyll synthesis, stem expansion, flower production and bud dormancy. Increasing the time the plants are exposed to light also allows the plant to make enough food to thrive.

- Humidity - High humidity causes slow usage of water by the plants which affect the quality of the crops. Low humidity causes the closing of the stomata to minimize transpiration thus reducing water loss. This however also reduces the rate of photosynthesis thus slowing plant growth. It is important to have anappropriate level of humidity.

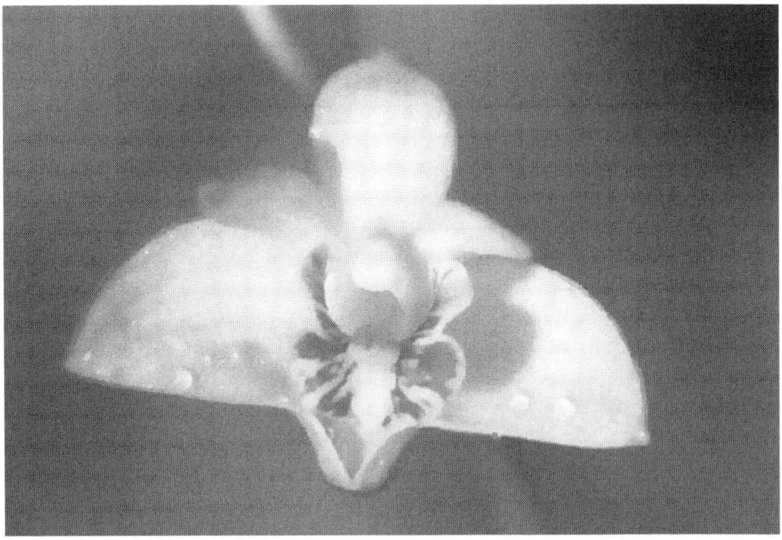

Figure 9.2: The relative humidity is affected by changes in temperature. A rise in temperature can cause condensation which wets the surfaces of the flowers and leaves.

Source: http://espacepourlavie.ca/en/relative-humidity-fertilization-and-flower-spike

- CO_2 - Adding CO_2levels during the flowering stage especially in the first two to three weeks promote flower production and also boost the size of the flowers.
- Pests - Proper spacing and eliminating weeds can help improve airflow and maintain appropriate levels of humidity thus helping avoid fungi and other pests and diseases. Overhead watering should be avoided to avoid increasing the humidity and to also avoid wetting the flowers and leaves. Watering should be done at the soil level.

9.3 HARVEST FACTORS AFFECTING FLOWER LONGEVITY

9.3.1 Time

Harvesting flowers too early can lead to harvesting immature flowers that are not developed sufficiently. These flowers may never bloom. Harvesting too late can lead to a reduction in the life of the flower. One needs to know the appropriate time to harvest flowers (Dahal, 2013).

Figure 9.3: This shows the difference in harvesting lavender to late and at the right time.

Source: https://getbusygardening.com/harvesting-lavender/

Depending on the flower, one has to know the point of maturity that each type of flower should be harvested. On top of that, the time of the day that the flower should be harvested is also important. Some examples and times the flower type should be harvested are as below:

- Marigold, Dahlia, Sunflower, Black-eyed Susan, etc. should be harvested once the flowers are fully opened.
- Columbine, Cockscomb, Daylily, etc. should be harvested when they are around 75% open.
- Late tulip, Bearded Iris, Alstroemeria, etc. should be harvested once the colored buds appear.

It is advised to harvest flowers in the coolest time of the day. Early mornings are not appropriate as the flowers are still coated by surface water from dew. During the rainy season, the flowers should be allowed to dry up in the afternoon sun before harvesting. Harvesting in the late morning or early evening is the most advisable time to harvest. This prevents soil-borne diseases such as powdery mildew and pests like aphids. Allowing them to get some sun before harvesting also allows the plant to store some carbohydrates which will act as a reserve for the flowering plant (Dahal, 2013).

9.3.2 Method of Harvesting

After the dew has dried, flowers can be harvested. They should be harvested in cool temperatures to ensure water levels are high. Cuts should be made to the plants to ensure long and well-built stems. Any foliage on the stems that are below the water should be removed to avoid the growth of bacteria and fungi when they decay.

Figure 9.4: Flowers should be harvested in small quantities at a time to avoid mechanical damage.

Source: https://www.johnnyseeds.com/growers-library/flowers/cut-flower-harvest-post-harvest-care.html

There are some flowers that love longer when all or most of the foliage is removed. Some flowers should also have their side shoots removed from the main stem. When harvesting, the cut on the stem can be slanted or straight. Slant cuts are advisable to avoid the stems from lying flat. This helps improve water intake. After harvesting, the flowers should immediately be graded and bunched. This protects them from any damage. If not immediately done, they should be kept in clean buckets with warm acidic water. The flowers should always be kept in a clean environment. Placing them on dirty surfaces can lead to contamination of the stems and that can lead to diseases. One needs to remember that harvesting is best done in small quantities at a time and the gardener should ensure that the flowers are immediately taken to a cool area to maintain their quality (Dahal, 2013).

9.3.3 Stage of Harvesting

Flowers need to be harvested at the right stage of development for long vase life. Each species has an optimum stage of harvesting. It is also influenced by the time of the year that the flower was planted. Some flowers should be harvested prematurely during summer to avoid fast maturity due to the warm temperatures. The general guideline states that spike type flowers should be harvested when ¼ to ½ of the florets are open while daisy type flowers should be harvested when they are fully open. Some examples are:

- Marigold – should be harvested when the flower is fully open.
- Roses – for the red and pink cultivars, they should be harvested when the first two petals begin to unfold and the calyx is reflexed below a horizontal position. Yellow cultivars should be harvested a bit earlier than the red and pink. White should be harvested later than the red and pink.

Figure 9.5: These roses should be harvested when the first two petals begin to unfold and the calyx is reflexed below a horizontal position.

Source: https://scvrs.homestead.com/BloomCycles.html

- Chrysanthemums – The standard cultivars should be harvested when the petals are fully enclosed. The spray cultivars should be harvested when the single flower opens but before anthesis.
- Gladiolus – these should be harvested when there are one to five buds showing color.

9.4 POST-HARVEST FACTORS AFFECTING FLOWER VASE LIFE

9.4.1 Flower Maturity

The minimum maturity a flower can be harvested refers to the stage at which the harvested buds can be opened fully and have a satisfactory display life after distribution. It is better to harvest at the bud stage then open them after storage, transport and distribution for most flowers.

Figure 9.6: Some flowers are harvested at the bud stage and others when they are fully open.

Source: https://pestid.msu.edu/weeds-and-plant-identification/british-elecampane-inula-britannica/

This process helps reduce the growing time for single harvest flowers. It also helps increase the packing density of the flowers. It simplifies the management of temperature and prevents the flowers from mechanical damage. Lastly, it prevents desiccation. Some flowers harvested at the bud stage are like roses and gladiolus but some are harvested when the flowers are fully opened like chrysanthemums and carnations. Flowers intended for the market are harvested when they are open compared to those that are being stored or shipped on long distances (Reid, 2009).

9.4.2 Temperature

Storage temperature is very important in improving the flower's vase life and quality of the flower. It is lowered from that of the field by a process called pre-cooling. If the storage temperature is high, the rate of respiration will increase and so will water loss. This will cause the flowers to wilt. Flowers should be cooled immediately after harvesting to ensure that they lose the heat from the field. They should be stored at the lowest temperature possible without freezing them or injuring them from chilling (Reid, 2009).

Table 9.1: These are some optimum temperatures for storage of different flowers

Flower species	**Common name**	**Storage temp. (°C)**	**The maximum period of storage**
Bellis perennis	English Daisy	4	3 days
Chrysanthemum frutescens	Marguerite Daisy	2	2 weeks
Cymbidium	Orchids	5-8	4 weeks
Dahlia hybida	Dahlia	4	3-5 weeks
Dianthus caryophyllus	Carnation (dry)	0-1	4-6 months
Dianthus caryophyllus	Carnation (wet)	4	4 weeks
Freesia refracta	Freesia	0-0.5	10-14 days
Gloriosa	Flame Lily	4-7	4-7 days
Narcissus	Daffodil	0-0.5	1-3 weeks
Rosa	Rose	0.5-3	2 weeks
Paeonia hybrid	Peony	0-7	4 weeks
Phlox subulata	Moss Phlox	4	1-3 days
Zantedeschia aethiopics	Calla Lily	4	1 week

Source: https://www.researchgate.net/figure/Showing-optimum-temperature-and-maximum-storage-periods-for-selected-flowers_tbl1_264837574

Most flowers last longer if kept in temperatures just near the freezing point but others may get injured in temperatures below 50 degrees F. Injury can be noticed by a darkening of the petals due to water soaking and even death of the flowers. Some of these temperatures are:

- Roses can be cooled at 1-3 degrees Celsius.
- Chrysanthemums can be cooled at 0.5-4 degrees Celsius.
- Carnations can be cooled at 1 degree Celsius.
- Gladiolus can be cooled at 4 degrees Celsius.

Pre-cooling can be achieved using different methods as given below:

- Room cooling – this involves placing the harvests in a refrigerated room or container. It uses little electricity but it is slow and mostly works for flowers that take longer to decay.
- Forced air cooling – this method uses fans that circulate cool air at a high speed in the storage room.
- Vacuum cooling – in this method, the air is drawn out of the chamber in which the flowers are stored. The pressure created allows the moisture in the flowers to evaporate.
- Ice cooling – this method is mainly used during transportation. Crushed ice is placed in a container or box and as it melts, the cool water lowers the temperature of the flowers.

9.4.3 Humidity

Humidity levels are normally quite difficult to control. The levels fluctuate due to air temperature changes. Plants also constantly transpire thus increasing water vapor in the air. Humidity in the air causes the plants to constantly change the stomatal openings of the leaves. High humidity causes slow usage of water by the plants which affect the quality of the crops. Low humidity causes the closing of the stomata to minimize transpiration thus reducing water loss.

The capacity of air to hold moisture increases with temperature. Water loss is directly proportional to the vapor pressure difference between the flower and its environment. The vapor pressure difference is inversely related to the relative humidity (RH) of the air surrounding the flower (Reid, 2009).

The relative humidity required by flowers is around 75-95%. High RH helps reduce transpiration allowing the flowers to keep well hydrated preventing them from drying out.

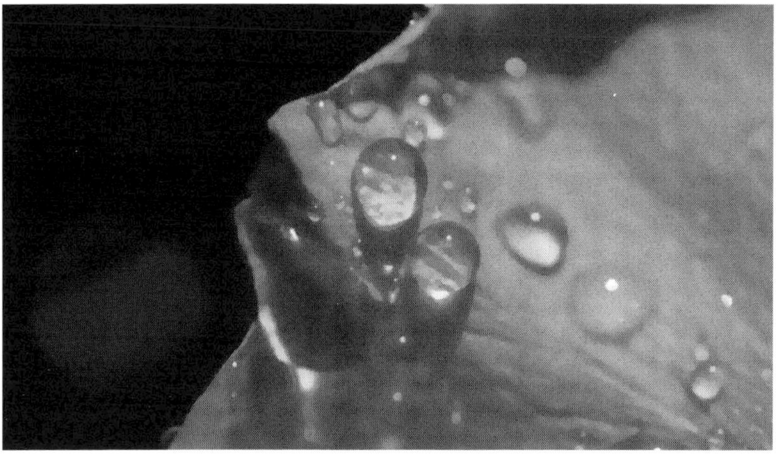

Figure 9.7: Condensation on flowers can lead to the growth of mold and fungi reducing flower quality.

Source: https://pixnio.com/nature-landscapes/water-dew-drops/flower-condensation-beautiful-dew-petal-moisture-pink-plant

Relative humidity can be controlled in a couple of ways listed here:

- Using humidifiers to add moisture in the surrounding.
- Ensuring that ventilation and air movement in the storage room is regulated in relation to the flower quantity.
- Wetting the storage room floors.
- Insulating the walls of the storage room using moisture barriers.
- Using polyethylene lining in the containers used for packaging.

Without the correct humidity, the flowers can lose water, decay, get certain diseases and even delay blooming of flowers that are harvested when they are not fully matured. Low humidity can cause higher levels of condensation which can also lead to decay of the flower petals. High relative humidity and free water encourage Botrytis while low relative humidity causes desiccation and browning of stems and pedicels affecting the quality and profitably of flowers.

9.4.4 Water Relations

The uptake ,transport, ability to retain water and water loss are all factors of water relations in the context of flowers. Harvesting separates the flower from its water source and from there the flower starts to lose water through transpiration. If water loss is extreme, it causes wilting or shriveling of the flower. Reducing water loss helps improve quality, longevity and profitability of the flower. One has to be careful because adding water directly to the flowers can have drastic consequences such as growth of powdery mildew and bacterial pathogens. Temperature changes after harvesting can cause condensation which can cause similar results. Therefore, there has to be a good balance between keeping water loss at a minimum and avoiding condensation post-harvest (Reid, 2009).

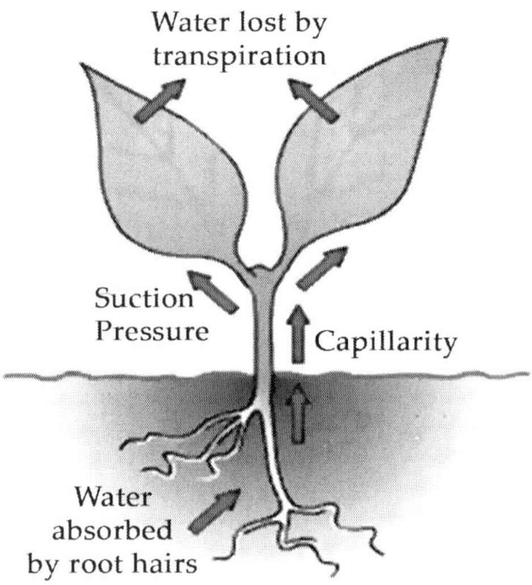

Figure 9.8: This shows how water is lost by plants through transpiration.

Source: https://www.vox4flowers.com/how-vox-works.html

Water loss is mainly caused by transpiration. The rate of transpiration is affected by the structure of the flower and environmental conditions. Temperature and relative humidity are major factors in water relations. Water-loss can be minimized by reducing the transpirational coefficient and by reducing the vapor pressure deficit. These can be done by keeping temperatures at a minimum, reducing the time between the harvest and the pre-cooling period, reducing air movement and increasing the relative humidity.

Condensation can be minimized by:

- Maintaining stable temperatures in the storage room.
- Maintaining a continuous cold chain
- Packing the flowers in a cold room.

9.4.5 Conditioning

Conditioning or hardening the flowers is a very important step in ensuring that cut flowers have a longer life. The stems should be recut while underwater to prevent air bubbles from forming in the stem. Two inches should be cut off from the bottom of each stem at an angle.

Flowers are conditioned by immersing them into tepid water that has preservatives specifically made for flowers almost up to the blooms. Plants with fuzzy foliage should not have their stems submerged while conditioning. On the contrary, they should have their foliage on the lower third removed and only immersed to that level. The containers in which the flowers are immersed should be placed in a cool area that is dark and has a high humidity for at least two hours or longer (Center for Agriculture, Food and the Environment, 2016).

The preservatives are used to supply sugar to the flowers that they need for proper growth and survival. They also act as a disinfectant preventing the growth of fungi or bacteria in the water. It comprises of a tablespoon of sugar and quarter teaspoon of bleach. The sugar is for nourishment and the bleach is to kill any bacteria and fungi. A quarter teaspoon of citric acid per gallon of water is also used as a preservative.

9.4.6 Floral Preservative

After harvesting flowers, they are not readily able to produce sugars that will help them survive. This can cause the flower to fade, lose its fragrance and have less strength. Floral preservatives help prevent these problems when added to vase water. It works best when mixed in warm water.

The contents change in different brands, but it has three key components:

- Sugar or dextrose
- An antibacterial agent
- An acidifier

The sugar or dextrose is used up by the flower as a source of energy to help the flower retain its fragrance and petal quality. This is important because photosynthesis does not occur once the flower is cut. The flower, therefore, needs a source of energy to survive longer.

Normally, the water has a pH above 5 making it alkaline. The flower sap is however at a pH of 3.5 which is acidic. For water intake to occur smoothly, the water needs to be acidified. Citric acid can be added to the water to lower the pH, making it easier for the flower to take in water (Center for Agriculture, Food and the Environment, 2016).

Figure 9.9: This is an example of a floral preservative. It contains sugar or dextrose, an antibacterial agent and an acidifier.

Source: https://www.amazon.com/slp/cut-flower-preservative/g9raoxodzce-g93j

Once a flower is cut, the bottom of the stem has an open wound that is susceptible to all kinds of bacteria and fungi. If the bacteria or fungi attack the open would, it may lead to the blockage of the stem, making it impossible for the stem to take in water or sugars from the vase. A biocide is thus incorporated in the preservative to kill any of these microbes. Common biocides used are chlorine or bromide. They are used in low quantities to avoid damage to the plant (Center for Agriculture, Food and the Environment, 2016).

9.4.7 Ethylene

Ethylene is the hormone responsible for flower and leaf aging. It is a plant stress hormone. Refrigeration is the most common way of preserving flowers. This method, however, leads to the accumulation of high levels of ethylene.

Some of the damages caused by Ethylene are as follows:

- It causes the flower to lose its shape
- It causes flower buds to drop prematurely
- It leads to the fading and loss of color of flower petals
- It also leads to the dropping of leaves and flower petals prematurely
- It causes the flower to take an abnormal shape as it opens

These effects can drop the profitability of the flower by a great deal. On top of that, it can lead to totally losing the product. During transportation, different flower species end up sharing the same space. This can cause major damage to the flowers because having a species that emits high levels of Ethylene next to one that does not emit much but is very sensitive to ethylene will cause damage (Center for Agriculture, Food and the Environment, 2016).

Figure 9.10: This is the effect of Ethylene on flowers. It causes fading and loss of color of flower petals.

Source: http://www.bioxtend.com/information/ethylene-facts

Technology has created ways to eliminate Ethylene from the air. Chemicals such as aminooxyacetic acid and aminoethoxyvinylglycin inhibit the biosynthesis of Ethylene. These are applied in aqueous solutions. There are also gaseous formulations that inhibit Ethylene action on flowers but are not efficient because the effect is not permanent. Silver thiosulfate can also inhibit Ethylene action on flowers but it is hazardous to the environment.

9.4.8 Growth Tropisms

This is plant growth away or towards a certain stimulus. This stimulus can either be water, gravity, light or even touch. Differential growth occurs when cells on one side of the plant grow faster than the other side of the same plant. This determines the direction the entire plant will grow in. If the plant is growing towards a stimulus, this is positive tropism. If the plant is growing away from the stimulus, it is negative tropism.

Figure 9.11: This is a positive tropism shown by the plants growing towards the light. Phototropism.

Source: https://www.thoughtco.com/plant-tropisms-4159843

Phototropism is when the plant grows towards the light. This tropism is important because plants require sunlight for photosynthesis. Potted plants should be kept close to a light source to promote photosynthesis. Chloroplasts are light sensitive and are the organelles that conduct photosynthesis.

Gravitropism is when the plant grows in response to gravity. This tropism is important as it allows the roots to grow towards gravitational pull and the stems grow against it. The root cap helps orient the roots inthe direction they should grow. This allows the plant to get water and nutrients from the soil, promoting proper growth of the plant (Gupta and Dubey, 2018).

Hydrotropism is when the plant grows towards the water. This tropism helps plants survive in drought conditions through positive tropism. It also helps prevent water saturation through negative tropism.

9.4.9 Mechanical Damage

Wounding the plant during harvesting, packaging or transport can lower the flower quality and life. It can also cause overproduction of Ethylene which will have many negative effects on the flowering plant. This wounding can happen if the flowers are not cut properly if they are not handled with care and even if they undergo over-application of chemicals.

Figure 9.12: Harvesting is best done in small quantities at a time to prevent mechanical damage by bruising.

Source: https://ag.umass.edu/greenhouse-floriculture/fact-sheets/harvesting-handling-cut-flowers

When harvesting it is advisable to use sharp scissors or a sharp knife so as to have a clean cut. Using a blunt knife will crush the stem and this will inhibit proper water intake. The harvested flowers should not be placed directly on the ground as that may cause the leaves and petals to be scratched or injured which can lead to infections with bacteria or fungi. The containers should not be overfilled with flowers as this can cause breaking of shoots and injury to leaves and petals due to brushing on each other and tangling.

The flowers should be graded and bunched immediately after harvesting them and before placing them in water so as to minimize contact with the flowers during the handling steps and this helps prevent mechanical damage. Mechanical damage to plants can greatly reduce the aesthetic value of flowers, so proper care and minimum handling areimportant (Gupta and Dubey, 2018).

9.4.10 Disease

Some of the bacteria that cause disease are Xanthomonas and Erwinia. Flowers are however mostly affected by fungi. Fungi cause root rot (Phytophthora), wilt (Fusarium), leaf and petal spots and flower blight (Botrytis). Botrytis blight is the most common diseaseof harvested flowers. This fungus causes flowers to blacken and decay. If the flowers are kept most for a long period of time (8-12 hours) they get a higher chance of getting Botrytis. Theselong periods of moisture give the fungus time to produce spores which are the infective stage (Dahal, 2013).

Figure 9.13: Botrytis blight is the most common disease of cut flowers.

Source: http://www.missouribotanicalgarden.org/gardens-gardening/your-garden/help-for-the-home-gardener/advice-tips-resources/pests-and-problems/diseases/fungal-spots/botrytis-blight.aspx

On top of that, flowers are stored at high levels of humidity which is around 93% or more. This also increases the chances of the flowers getting infected by the fungus especially in temperatures of 55 to 65 degrees F. Botrytis can be prevented in a couple of ways:

- One should not harvest and store wet flowers.
- Flowers should not be moved from cold rooms to warm rooms immediately. If they are, it will cause the formation of droplets on the petals and leaves due to condensation.
- The floors in the processing rooms should be dry so as to avoid increasing the relative humidity.

9.5 HARVESTING FLOWERS

Harvesting flowers requires a well thought out process. Here is the process:

- One should have enough clean buckets before cutting flowers. The buckets and pruners should be well cleaned with a tablespoon of bleach per gallon of water. They should soak for a while then rinsed off with clean water. Thereafter, the bucket should be filled with a third with clean water. These steps are important in preventing diseases.
- The buckets should be kept in a shade which can allow the flowers to cool in the buckets immediately after cutting. To cut the flowers, one should use a sharp knife or scissors. This will avoid crushing the stems. Any foliage below the water should be removed and any dirt shaken off. This is done to ensure the bucket is clean at all times.
- The flowers should be harvested when they are dry, cool and well hydrated. This is either in the late morning or early evening. Harvesting wet flowers will cause the growth of fungi and mold.
- After harvesting, the buckets should be placed in a cool area. The stems should be recut underwater in a slant manner so as to avoid water bubbles from entering the stem.
- Thereafter, the flowers can be placed in a clean bucket with clean water that has floral preservatives. After this, the flowers can be arranged.

While harvesting, one needs to know all the details of the specific flower being harvested. Each species has a certain level of maturity it can be harvested at. Marigold, Dahlia, Sunflower, Black-eyed Susan, etc. should be harvested once the flowers are fully opened. Columbine, Cockscomb, Daylily, etc. should be harvested when they are around 75% open. Late tulip, Bearded Iris, Alstroemeria, etc. should be harvested once the colored buds appear (Center for Agriculture, Food and the Environment, 2016).

Figure 9.14: One should have enough clean buckets before cutting flowers. The buckets and pruners should be well cleaned with a tablespoon of bleach per gallon of water.

Source: https://thenerdyfarmwife.com/how-to-harvest-dry-flowers-herbs-from-garden/

9.6 FLOWER STORAGE METHODS

9.6.1 Simple Refrigerated Storage

This is the most common method of flower storage. Most flowers prosper at temperatures of 32 to 33.8 degrees F and relative humidity of 95 to 99%.

Flowers can dry out quickly due to their high respiration rate. They need to be cooled as soon as possible after harvesting. This will allow them to live longer.

Figure 9.15: Refrigerators must have precise temperatures for the flowers they are storing to ensure maximum flower quality. The ideal temperature for a flower refrigerator is between 1.11 to 2.22 degrees Celsius.

Source: https://www.uscooler.com/products/applications/floral/

Some flowers like tropical flowers are sensitive to low temperatures. They should not be stored in temperatures below 50 degrees F. Lower temperatures cause them to have darkened petals and even dry out, e.g. Orchids.

Tulips prefer temperatures of 33 to 35 degrees F and an RH of 90% and above. Their shelf life is approximately three to seven days. Roses, on the other hand, need to be stored at 33 to 35 degrees F to ensure blooming.

There are two methods of storage: wet storage and dry storage. Wet storage involves flowers that are stored with the stems immersed in water or a preservative. This is a good method if the storage time is short. It works best at 2-4 degrees Celsius. Dry storage involves sealing the flowers in plastic bags before storage to prevent loss of moisture. This is appropriate for long storage periods. One needs to pre-cool and pulse before using this method. It works at temperatures of 0.5-1 degree Celsius (Center for Agriculture, Food and the Environment, 2016).

Refrigerators must have precise temperatures for the flowers they are storing to ensure maximum flower quality. The ideal temperature for a flower refrigerator is between 1.11 to 2.22 degrees Celsius. These temperatures are ideal for cut flowers that are in a hydrating solution. Flowers that are not in a hydrating solution require a temperature of 34 degrees F. Many refrigerators maintain a temperature of 35 degrees F. Any fluctuations plus or minus five degrees can reduce the life and quality of the flowers.

On top of the required temperature, the humidity levels inside the refrigerators need to be precise as well. They require humidity levels of 90 to 95%. If the flowers are placed in a preservative, they can thrive in any humidity level above 80% (Center for Agriculture, Food and the Environment, 2016).

9.6.2 Controlled Atmosphere Storage

This storage method works by decreasing the level of oxygen and increasing the concentration of CO_2 by controlling the gas composition precisely. For this method to work, the gas atmosphere needs to be controlled continuously throughout the storage period. A device called a scrubber is used to absorb any excess gas in the packaging material. Air contains 21% of Oxygen and 0.03% of Carbon dioxide. Increasing the concentration of Carbon dioxide and lowering the level of oxygen can help improve the vase life of the flowers (Gupta and Dubey, 2018).

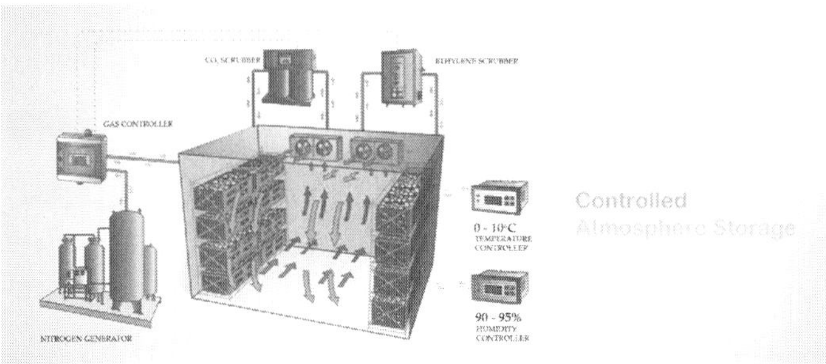

Figure 9.16: This storage method works by decreasing the level of oxygen and increasing the concentration of CO_2 by controlling the gas composition precisely.

Source: http://www.agroripe.com/controlled-atmosphere-storage/

Advantages of this method are:

- It delays the softening of petals and leaves
- It delays the hardening of the stems
- It promotes the retention of chlorophyll
- It helps control any insects or pests due to low levels of oxygen
- It prevents browning of the stems on the cut surface
- It retains the fragrance of flowers
- It reduced the chances of fungal growth
- High levels of CO_2 allow the flower to retain more nutrients for better flower quality

Disadvantages of this method are:

- It may cause the flowers to develop too quickly thus shortening the vase life
- It may cause some physiological disorders
- It may enhance anaerobic respiration
- It may cause the flowers to decay
- It is an expensive method to maintain
- It requires the control of temperature as well
- If the pack leaks or opens, ask the benefits are lost

In conclusion, this method can better the quality of the flowers after storage and can allow flowers to be stored for a longer proud of time compared to refrigeration. It allows the flower to have a longer vase life after storage. On top of all these, this method is environmentally friendly because with it you do not need post-harvest chemical treatments if the flowers (Center for Agriculture, Food and the Environment, 2016).

9.6.3 Hypobaric Storage

This is also called the lower pressure storage which refers to storing flowers at reduced atmospheric pressure. This method done under refrigeration lowers the respiration rate and helps evacuate Ethylene. Storing flowers at atmospheric pressure of 0.2-0.5 at 15-24 degrees Celsius causes a reduction of ethylene and oxygen in the chamber. Lower oxygen levels reduce respiration rates and production of ethylene.

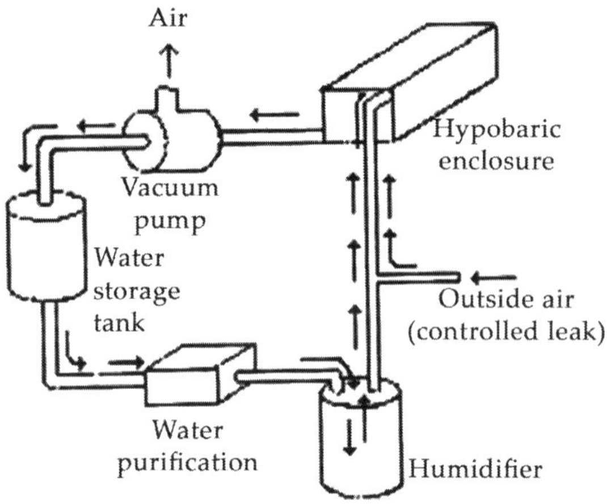

Figure 9.17: This method has two disadvantages. The cost of installation is too high and it causes water loss from the tissues.

Source: https://link.springer.com/chapter/10.1007/978-3-319-23591-2_3

The gas composition depends on the amount of product, the respiration rate, the respiration quotient and the exchange between the storage room and the surroundings as well as the storage room and the scrubber.

There are two main disadvantaged of this method. The cost of installation is too high and it causes water loss from the tissues. Each flower has a specific storage temperature and a specific maximum time of storage:

- Carnations stored dry require temperatures of 0-1 degrees Celsius for a maximum of 16-24 days. When stored wet they require temperatures of 0.5-1 degrees Celsius for a maximum of 21-28 days.
- Roses stored dry require temperatures of 0.5-2 degrees Celsius for a maximum of 7 days. When stored wet they require temperatures of 2-3 degrees Celsius for a maximum of 5-7 days.

9.7 STORAGE CONDITIONS FOR FLOWER CROPS

9.7.1 Temperature

Temperature controls the timing of the crops. It dictates the time between the planting of the crops to the flowering of the plant. Plants grow flowers faster when grown in warm areas (an increase in temperature increases the plant development rate). Warm temperatures promote chlorophyll synthesis and development of the leaf surface area (FAO, 2013).

Storage, however, is different, as the flowers need to be kept in cool areas instead. If the storage temperature is high, the rate of respiration will increase and so will water loss. This will cause the flowers to wilt. Flowers should be cooled immediately after harvesting to ensure that they lose the heat from the field. They should be stored at the lowest temperature possible without freezing them or injuring them from chilling. Most flowers last longer if kept in temperatures just near the freezing point but others may get injured in temperatures below 50 degrees F. Injury can be noticed by darkening of the petals due to water soaking and even death of the flowers (FAO, 2013).

Figure 9.18: Flowers need to be kept in cool areas instead. If the storage temperature is high, the rate of respiration will increase and so will water loss.

Source: https://www.marjoriemilner.edu.au/florist-cool-room/

Storing flowers at warm temperatures causes condensation. This makes the surface of the petals and leaves wet which can cause the growth of mold and fungi. Condensation can be minimized by the following:

- Maintaining stable temperatures in the storage room.
- Maintaining a continuous cold chain
- Packing the flowers in a cold room.

9.7.2 Relative Humidity

Humidity in a storage room is hard to control. Without the correct humidity, the flowers can lose water, decay, get certain diseases and even delay blooming of flowers that are harvested when they are not fully matured. Low humidity can cause higher levels of condensation which can also lead to decay of the flower petals. Water loss of 3 to 6% is generally enough to cause a noticeable loss of quality and value thus reducing profitability.

Figure 9.18: The relative humidity required by flowers is around 75-95%.

Source: https://www.rimolgreenhouses.com/blog/controlling-greenhouse-humidity-during-summer-s-hottest-months

High relative humidity and free water encourage Botrytis while low relative humidity causes desiccation and browning of stems and pedicels affecting the quality and profitably of flowers. The relative humidity required by flowers is around 75-95%. High RH helps reduce transpiration allowing the flowers to keep well hydrated preventing them from drying out (Reid, 2009). Relative humidity in a storage room can be increased by adding moisture to the air by using sprays, humidifiers or by wetting the floor. Vapor barriers can also be used. Examples of these are waxes or using polyethylene liners in boxes. Ensuring that ventilation and air movement in the storage room is regulated in relation to the flower quantity can also help maintain the relative humidity.

9.8 GRADING AND SORTING FLOWERS

Cut flowers are sold by grades. Carnations, for example, have six different grades: shorts, splits, heads, standard, selects and fancy. They differentiate more profitable cuts from others. Selects are the most expensive followed by fancy then standard then shorts. These take much more time to process than the other grades.

Figure 9.20: There are grading stations with a grader at each. The grader takes the ungraded cut flower material and grades and simultaneously bunches the cut flower material in fixed, decision limiting racks.

Source: http://indifresh.co.uk/about.html

There is a central grading system that eliminates and minimizes problems such as different gradings between different growers and too much handling that may cause mechanical damage to the plant. The central grading system provides consistent products which improve the reward quality. This helps achieve economies of scale.

There are grading stations with a grader at each. The grader takes the ungraded cut flower material and grades and simultaneously bunches the cut flower material in fixed, decision limiting racks. Each rack corresponds to a different grade of flower (Reid, 2009).

Sorting flowers has a criterion of its own:

- One should follow the local and export market requirements when sorting flowers.

- The stem thickness needs to be uniform in one bunch.

Sorting is done by hand to remove any unwanted parts on the flower that are unsuitable for the market such as damaged parts or ill parts. They are then separated on the basis of color, shape or visible defects.

9.9 BUNCHING

All flowers apart from anthuriums, orchids and a few others are normally bunched. The number of flowers in a bunch varies depending on the species and market of the flower. Single stemmed flowers come in groups of ten, twelve and twenty-five.

Figure 9.21: Strings or elastic bands are used to hold together the bunches.

Source: https://felins.com/media-gallery?related_industries=All&media_type=All&field_media_product_technology_target_id=All&page=50

Spray type flowers are bunched according to the number of flowers that are open, the size of the bunch or the weight of the bunch. Strings or elastic bands are used to hold together the bunches. Sleeving is done as soon as possible after the harvest for specific reasons cited below:

- To avoid tangling of the flowers.
- To separate the flowers.
- To protect the flower heads from damage.

Sleeving is done using waxed or unwaxed paper, a card that is corrugated but with the smooth sides towards the flowers and polyethylene. All these processes should be done in the field or greenhouse to avoid too much handling which can cause mechanical damage to the flowers.

9.10 CHEMICAL SOLUTIONS

9.10.1 Rehydration

This is used after harvesting to improve the quality of the cut flower. If the flowers are wilted, they require rehydration to restore their turgidity. The water should be deionized and it should have a biocide. The water should be acidic at a pH of around 3.5. This can be done using citric acid or aluminum sulfate. During rehydration, no sugars should be added. The temperatures should be low thus it should be done in a cooler.

Figure 9.22: This shows the results of the rehydration of flowers.

Source: http://blog.floralife.com/2016/05/23/floral-research-water-uptake-fresh-weight-vase-life/

One should replace the solution often so as to avoid a cloudy solution. Cloudy solutions are a sign of high microbe numbers. If the water has microbes, they will travel up the stem and cause blockage, preventing proper uptake of water by the plant. Tap water should be avoided because of the ions in it. The time a flower takes to rehydrate depends on the particular variety of flowers. While rehydrating, the stems should be cut at an angle underwater to prevent water bubbles. At least two inches should be cut off. This opens up the water channels. The flowers should sit in the solution for a minimum of two hours (Center for Agriculture, Food and the Environment, 2016).

9.10.2 Pulsing

Pulsing refers to placing newly harvested flowers in a solution for a few seconds to several hours so as to ensure they maximize their storage and lifetime. Every different crop has a specific solution. It has three main functions:

- It provides sugars to the flower such as Gladiolus, Hybrid Statice and Lisianthus
- It extends the life of flowers that are very sensitive to ethylene such as carnations
- It prevents the leaves from yellowing in plants like Alstroemeria

The main ingredient of pulsing solutions is sucrose. It is available in the solution in a concentration of two to twenty percent depending on the crop. It helps provide additional sugar. On top of that, it should always have an additional biocide that can work with the specific crop being treated. For flowers that are sensitive to Ethylene, they should be pulsed using Silver Thiosulphate. This should be done at temperatures of 20 degrees Celsius for 10 min if pulsing for a short period, or 2 degrees Celsius for 20 hours if pulsing for a long time. Flowers like Alstroemeria gibberellic acid can help in preventing the yellowing of the leaves. Some flowers prefer short pulses in silver nitrate whose action on the plant is not fully understood but has shown properties of acting as a biocide especially in chrysanthemums but it should be rinsed off before packing (Dahal, 2013).

9.10.3 Bud Opening

Before flowers are sold to the consumer, they should be opened in bud opening solutions. These solutions have biocide and sugar. These solutions have proven to improve the quality of flowers compared to when they open in the environment.

Figure 9.23: Budding solution improves the balance of water and also improves the longevity of the flowers.

Source: https://www.shutterstock.com/search/flower+bud+opening

Quaternary ammonium salts have a strong biocide power and have a good effect on the opening and conserving of certain flowers such as carnations. It is formulated to have sucrose, citric acid and benzalkonium chloride. It has proven to improve the speed of opening, flavor, longevity and corolladiameter of certain flowers in relation to flowers that have opened naturally. Some advantages of this solution are:

- It is economically available.
- It has no precipitates making it colorless and transparent.
- It does not require an addition todistilled water.
- It does not alter the stems of the flowers and it is not phytotoxic.

Other solutions have silver thiosulfate instead of benzalkonium chloride in the solution. This solution increases bud opening and inflorescence longevity. It can be used 24-48 hours before transportation. It has 10-15% sucrose and 0.25mM of Silver thiosulfate. This solution improves the balance of water and also improves the longevity of the flowers.

9.11 PACKING

9.11.1 Box Design and Construction

Packing boxes come in many shapes and sizes. Most come long and flat which reduces the depth of the flowers in the box. This can prevent mechanical damage tothe flowers. There are many different box designs all over the earth. The material used to make the boxes varies. Cardboard can become very inconvenient in areas with high humidity due to loss of strength that may cause the box to collapse. The material needs to be of good quality to avoid this. A white card is of high quality and it also allows the provider to print good designs on the box.

Figure 9.24: Cardboard can become very inconvenient in areas with high humidity due to loss of strength that may cause the box to collapse. One needs to ensure the box material is of good quality.

Source: https://www.pinterest.com/pin/687503317836195931/

Boxes can have some flaws. One of the main ones found in some boxes is pre-forming of the pre-cooling vents which are not sealed. When these boxes are handled at high temperatures, they create a warm environment for the flowers which can lead to condensation in the box. One of the boxes called the California type box has a flap system that is sealable. It is appropriate for flowers that are packaged and transported in a warm climate. The flaps are sealed after pre-cooling allowing the flowers to avoid the external temperatures. Flaps on the box provide extra strength to the corners. If using staples, they should be enough to make sure that the cardboard is stiff. If you can't use staples since some marketing chains do not approve it, hot glue can be a replacement. To add strength to the box, one can also form a triangular strut in the bottom corners of the box. The air gap between the flowers and the box provides more packing volume so the strut does not affect that (Reid, 2009).

9.11.2 Packing Systems

Air transport is quite expensive. This has led to more people using packing systems. These systems can greatly reduce the quality of the flower. The producers tend to overpack the boxes because the boxes are measured by volume, not by weight. This leads to a box that is rounded and cannot stack well making the product carry the load. This affects pre-cooling and the box can easily fail.

Figure 9.25: Overpacking boxes should be avoided so as to maintain the pre-cooling effect.

Source: http://packagingsystemsincorporated.com/

For the flowers to be of good quality, they require boxes that protect them and overpacked boxes cannot provide protection to the flowers. The forced air pre-cooling system depends on 5 to 8 cm plenum on each side of the box so that cool air is distributed to all the flowers. When the boxes are over packed, it leads to warming and damaging the flowers which will then have a low vase life and low aesthetic value.

9.12 COOLING

9.12.1 Cooling Time Calculations

A cooling curve expresses the time required to reach a certain temperature. The time that is required to reduce the temperature of the flowers from the initial temperature to the temperature of the air that is cooling is a commercial cooler target.

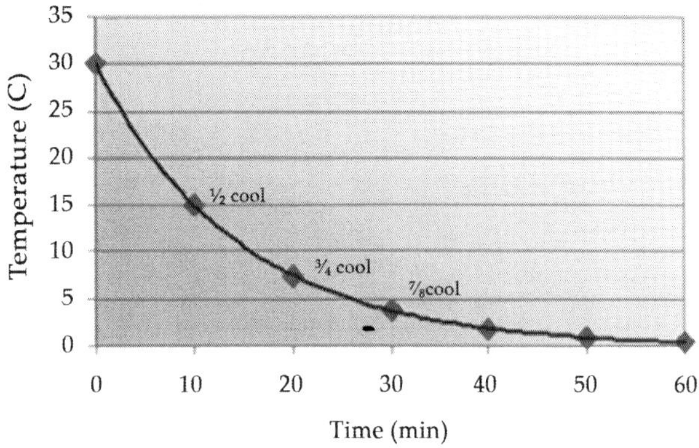

Cooling curve for cut flowers in a forced air cooler with supply air at 0 C. The half-cooling time for these flowers is 10 minutes

Figure 9.26: According to the diagram, cooling would be dome for more than two hours to reach the temperature of the cooling air.

Source: http://ucce.ucdavis.edu/files/datastore/234-1906.pdf

The rate of cooling slows down as the flower's temperature reaches that of the refrigerated air. The flower may not reach the exact temperature of the refrigerated air. According to the diagram, cooling would be dome for more than two hours to reach the temperature of the cooling air. The cooling air should be able to flow through the box without being blocked by any material. This means that less packing paper should be used in packing flowers meant for pre-cooling.

9.12.2 Fans

Forced air cooler uses centrifugal or axial flow fans. The latter ismuch louder than the former. Centrifugal fans can force more air against higher static pressure but may require more horsepower than axial flow fans to operate. Two criteria are used to select fans:

- The airflow that is required measured in cubic feet per minute
- The static pressure that is required measured in inches of water

These two are dependent on the flower type, the cooling rate that is desired and the number of boxes being cooled. A pre-cooler can be designed and requires estimations of the airflow required. This can be achieved by multiplying the required L/sec per box by the number of boxed that need to be cooled (based on a peak day) and adding 25% to account for any air leaks. Boxes stacked end to end should not be cooled (Center for Agriculture, Food and the Environment, 2016).

If the packages are situated in a cool environment, the vents can be left open to allow the escape of heat of respiration. If the packages are being transported in warm climate, the vents should be resealed. Any ice used after pre-cooling is only useful if it is surrounding the product, therefore, preventing entry of heat into the carton. This should be done carefully to prevent melting of the ice on the flowers of causing damage by freezing.

9.12.3 Tropical Flowers

Tropical flowers need to be packed in enough insulating material especially if packed in a mixed load. An example is an insulated box packed with shredded newsprint. Tropical flowers should not be pre- cooled. If they are being transported by a truck that is refrigerated, they should be placed at the center of the rest of the load. This is to prevent any cooling air from reaching the package with tropical flowers.

Figure 9.27: These are examples of tropical flowers thatare of high market value.

Source: https://www.thespruce.com/tropical-flowers-you-can-grow-any-where-4122087

Tropical flowers have a high market value, therefore, should be transported by air because they are easily damaged by low temperatures.

9.12.4 Cooling for Aqua Packs

This refers to packing flowers vertically in boxes that have water with a biocide in the bottom. For air transport, one has to consider the weight and the possibility of spillage, therefore, they should be packed without any solution. The solution can be added after transportation.

Figure 9.28: These flowers are packed vertically in boxes that have water with a biocide in the bottom.

Source: https://personalflowersofcrewe.co.uk/en/home/99-diamond-wedding-aqua-pack.html

These boxes can be a bit difficult to pre-cool. One can place palletized aqua packs on a modified plenum that is designed to draw air down through the flowers and out through the hamper bottoms which contain ventilation grids. Padding above the air gap can help seal air from the bottom of the boxes and the pallets. The air is drawn down the boxes. This process is ensured by a strip of canvas or sheet of fiberboard that can be placed at the bottom of the boxes and the pallet. Therefore, precooling these boxes is achieved on a horizontal conventional pre-cooler by directing the airflow down through the boxes withthe help of an adapter plenum and a fabric sleeve around the base of the boxes (Gupta and Dubey, 2018).

9.12.5 Vacuum Cooling

This method is rarely used commercially. Recently, vacuum coolers have been installed in places where flowers are marketed such as near airports but some growers have also incorporated them so as to remove heat from packed flowers. This method works on the basis of water cools at a lower temperature when boiled at low atmospheric pressure. If water is boiled at a tenth of the atmospheric pressure, it reaches boiling point at 0 degrees Celsius. When flowers are placed in this low-pressure level, the water in their cells boils thus releasing heat (Gupta and Dubey, 2018).

Advantages of this method are given as follows:

- The efficiency of cooling is not affected by any paper or plastic packaging material.
- It eliminates free water.
- The boxes do not have to be stacked in a specific way in the pre-cooler.
- It works equally well with dry packs or wet packs for cut flowers.

Figure 9.29: This method works on the basis of water cools at a lower temperature when boiled at low atmospheric pressure.

Source: https://webercooling.com/en/flowers-cold-chain-management

This method has one main disadvantage. The equipment has a high capital cost. If the facility is centralized, the high cost can be catered for by the high throughput of boxes that this equipment can achieve.

9.13 IRRADIATION

Any flowers shipped across international borders are subject to inspection for any diseases or pests that might be harmful. For a couple of years, it has been suggested that one can sterilize flowers and/or kill insect pests without damaging the flowers by using ionizing irradiation. It has been suggested further that not only will it sterilize but also improve the longevity of the flowers.

Figure 9.30: This process is done by passing packed flowers through a radiation chamber on a conveyor belt.

Source: https://www.researchgate.net/figure/Red-ginger-flowers-18-d-post-irradiation-left-335-422-Gy-vs-flowers-not-irradiated_fig1_263566024

Ionizing irradiation once performed leaves no residue. Doses as low as 0.2 Gy can kill insect pests. This process is done by passing packed flowers through a radiation chamber on a conveyor belt. It is passed through a radiation beam that is like a large flashlight to avoid coming in contact with the radioactive material. This energy breaks the bonds in the microorganism's DNA molecules making the microorganism die or infertile (Reid, 2009).

There are four sources of ionizing radiation:

- Cobalt-60 gamma sources
- X-ray accelerators

- Electron beam generators
- Gamma rays

Insects pests with large quantities of DNA are killed rapidly by a low dose of radiation. Bacteria are harder to kill because they have lesser DNA. Viruses are resistant to irradiation at doses acceptable for flowers. This method does not change the nutritional value of the flowers.

10

SUSTAINABILITY IN FLOWER CROP CULTIVATION

CONTENTS

10.1 Introduction 262
10.2 Factors Affecting Sustainability In Flower Crop Cultivation............ 262
10.3 Benefits Of Sustainable Flower Crop Cultivation 267
10.4 Initiatives To Maintaining Sustainability In Flower Cultivation. 273

10.1 INTRODUCTION

Sustainability is the ability to meet the present needs (in this case, of flower cultivation) without compromising the ability to meet those in the future. This concept has three pillars: economic, social and environmental. Economic sustainability is very important in flower crop cultivation. The population is currently growing, making it harder and harder to provide for the whole population. On top of that, the environment is slowly degrading and climate changes are becoming very unpredictable. Gardeners have to adapt their cultivation system to this. To be able to sustain the business economically, one needs to consider and comply with the market demands and trend or all in all find a whole new market. There are many things that should be done to maintain sustainability. As gardeners, one needs to ensure that they are monitoring and adjusting farming practices so as to improve the soil and environment around it. Fertile soil is productive and clean water is more affordable and better for floricultural use thus complying with market demands and providing economic and social gain.

To ensure that the project will be feasible, the farmer needs to tick a few boxes before starting any farming project:

- What is the climate in the area?
- What is soil depth?
- What is soil mapping?
- What is the soil type?
- Are there any contours?
- One has to do soil and water testing.

To make it possible to maintain sustainable production, the soil and water should be tested annually. At the very least, a root inspection, nematode analysis and soil analysis should be done.

10.2 FACTORS AFFECTING SUSTAINABILITY IN FLOWER CROP CULTIVATION.

10.2.1 Nutrients

The provision of nutrients to flower crops is important and it promotes a good yield of high-quality flowers for sales. This should be achieved without contaminating the environment and maintaining the soil's health by using fertilizers in an appropriate manner (Rikken & ProVerde, 2010).

Some of the major nutrients are:

Nitrogen – Low nitrogen levels in the soil result in lower flower yield, poor plant growth and the formation of foliar chlorosis. Excessive nitrogen can, however, cause formation of weak stems, delayed flowering and excessive vegetative growth. Nitrogen leaches from the soil as nitrates which can pollute water bodies. Nitrogen also influences soil pH. Ammonium fertilizers can cause soil acidity when the ammonium is converted to NO_3 by bacteria in the soil. Lime can be used to neutralize the acidity.

Phosphorus – Low levels of phosphorus may cause the plant to be small in stature and also produce short stems. It also promotes the flowering of plants. Phosphorus, however, causes negative effects onthe environment by running off to water bodies causing the growth of algae which lowers the water quality.

Potassium – Low potassium levels can result in marginal chlorosis and the burning of the leaves on the lower part of the plant. Excessive deficiency may affect the leaves on the flowering stems and reduce the yield. Low potassium levels are mainly seen in soils with high sand levels and those with little to no organic matter. Soils having little to no organic matter have low cation exchange capacity and can therefore not retain positively charged ions such as potassium, magnesium and calcium (Rikken & ProVerde, 2010).

Figure 10.1: Soil nutrients are important for flower crops cultivation.

Source: https://www.gettyimages.ae/detail/photo/miniature-tools-for-floriculture-small-shovels-royalty-free-image/859711240

If necessary, limestone can be added to the soil. It is important to maintain soil pH at 5.5 to 7 so as to maintain soil fertility. The pH affects the presence of trace elements such as phosphorus which is highly available at a pH between 6 and 7. Limestone can also act as a fertilizer for magnesium and calcium (calcitic limestone with 5-10% magnesium). Liming also prevents the buildup of aluminum and manganese which can be toxic. Earthworms can be a good sign of fertile soil.

10.2.2 Soil

The soil has major functions for plant growth as follows:

- It assists in the anchoring of the plant to the ground.
- It keeps moisture and provides water to the crops.
- It is a source of nutrients to the plant.

The soil has 4 components. It has water, air, organic matter and minerals. The soil has three functional phases. The liquid phase has hygroscopic water which is bound and not readily available to the plants, capillary water which is held by the capillaries and is available to plants and gravitational water which is held by tension and provides a lot of water to plants. The gas-phase contains air pockets with Nitrogen, oxygen and CO_2(De Pascale and Maggio, 2005).

Proper management of the soil allows the soil to be at its peak performance improving the quality of the flowers. One should analyze the soil and check for the amount of humus in the soil, the value of pH and the amount of nutrients in the soil.

Soils also require enough organic matter which will improve the soil structure and improve the ability of the soil to hold water and nutrients. It also prevents soil erosion and compaction assisting in promoting soil organisms. Good soil drainage is also important. If the soil does not drain water the soil becomes a dump and that is a good breeding place for fungi and other diseases.

10.2.3 Land

The quality of the land and that of the soil are related to land sustainability. For land to be sustainable, there are a few factors that need to be considered:

The land-use intensity – this describes how the intensification of agriculture on the land will impact land quality. This involves increasing the number of crops planted on the land, increased frequency of planting and value-added production. These transitions to intensification need to be well managed by the farmers to ensure that the land is well accounted for.

The land cover – this describes the extent, timing and duration of the flower crop cover on the land during periods that can cause major soil erosion on the land. This gives a better understanding ofdesertification.

The yield trends and yield gaps – this describes the actual potential of the land to grow the flower crops during the trends as expected by the market.

Land degradation – this involves organic matter loss, erosion and compaction.

Soil quality – this is based on the turnover of soil organic matter. In particular, the microbiological carbon pool affected by environmental conditions and land use.

Land contamination and pollution – this is based on whether the land is exposed to any pollutants that may lower the flower crop quality.

All these are indicators of land sustainability that need to be considered before starting any flower cultivation project.

10.2.4 Energy

Flowers are grown all year long. Some farmers use farm machines that are driven by fossil fuels which lead to greenhouse gas emissions which cause climate change leading to environmental damage. On top of this, some farmers replenish the soil using natural gas to convert Nitrogen in the air into synthetic fertilizer. Natural gas is non-renewable. Some use phosphate or fossil fuel in diesel generators to pump water for irrigation. All these sources of energy are non-renewable (De Pascale and Maggio, 2005).

Most non-renewable sources of energy are fossil fuels such as petroleum and natural gas with carbon being their main element. They have some advantages such as they are inexpensive to extract and they can be stored, shipped and transported anywhere. However, the disadvantages outweigh the advantages. Some of the disadvantages are:

They contribute to global warming – since they are not green sources of energy they contain large amounts of carbon. Burning them produces carbon dioxide which is a high contributor to global warming. This can also cause acid rain which will not be good for the flower crops.

They are non-renewable – this means that that fossil fuels are available in high quantities but the reserves are not replenished naturally. In comparison to solar energy and wind energy, fossil fuels take years to form in any noteworthy quantities.

They are unsustainable – we spend the reserves in a manner that cannot be sustained meaning we have to look for other sources of energy.

10.2.5 Water

Water is the largest natural resource yet only 3% of it is fresh. Of the 3%, only a third of it is accessible for use in agriculture and in the cities. The rest of it is frozen or too deep in the ground to retrieve. Agriculture in general uses about 70% of the yearly water consumption globally. An increase in demand will greatly affect the agriculture sector.

Figure 10.2: Investing in irrigation methods that are low pressure and low volumes such as drip irrigation and the use of micro-sprayers can help minimize water use.

Source: https://ec.europa.eu/jrc/en/news/minimum-quality-requirements-water-reuse-agriculture

As stakeholders in the agricultural sector, flower crops farmers should use different techniques in water use and management such as:

- Farmers should invest in techniques and technology that use water efficiently such as allocating water in better ways and improving irrigation systems.
- Using techniques such as nutrient management and conservation

tillage to retain the moisture of the soil, prevent soil erosion and build up soil organic matter.

- Being well prepared for droughts by using weather forecasts, planting drought-tolerant crops, using evapo-transpiration as a method of improved irrigation.
- Building appropriate and efficient irrigation systems.
- Building water storage equipment.
- Water loss can also be minimized by avoiding pre and post-harvest losses.
- Finding ways to recycle and treat wastewater to irrigate some crops.
- Investing in irrigation methods that are low pressure and low volumes such as drip irrigation and the use of micro-sprayers. Any leaks in the irrigation channels should also be fixed.
- Using natural aquifers is more beneficial than using dams to store water because it prevents water loss through evaporation.

On top of these, watersheds, wetlands and pasturelands should be protected to preserve the ecosystem.

10.3 BENEFITS OF SUSTAINABLE FLOWER CROP CULTIVATION

10.3.1 Environment Preservation

Preserving the environment can protect the soil and make flower cultivation more sustainable. It leads to improved water quality, improved air quality, carbon sequestration, reduced soil erosion and an increase in biodiversity. These can be achieved by practicing conservation agriculture. This technique has three main principles:

- The soil should be disturbed as little as possible by avoiding hoeing and plowing.
- The soil should be covered as much as possible by using mulch, cover crops or crop residues.
- The crops should be mixed and rotated which can help maintain soil fertility and minimize pests and diseases.

Figure 10.3: Destroying soil cover can lead to a rise in pests and disappearance of beneficial animals, insects and micro-organisms thus lowering the quality of flower crops.

Source: https://greentumble.com/10-sustainable-farming-methods-and-practices/

Crop residues reduce the splash effect when it rains. Reducing the energy of the raindrops they then proceed to the soil without affecting the soil negatively. This reduces runoff and it leads to higher infiltration. These two effects can reduce soil erosion greatly. The residues also act as a physical barrier that can reduce the speed of water and wind on the surface of the soil. Lower wind speed reduces evaporation of soil moisture.

Conserving the soil cover creates a habitat for different species that feed on pests thatattract more birds and insects that help in pollinating the flowers. In contrast, destroying soil cover can lead to a rise in pests and disappearance of beneficial animals, insects and micro-organisms (Ikerd, 2003).

10.3.2 Economic Profitability

Economic sustainability requires a farm to be managed in a manner that ensures long term profitability. Farm revenues are volatile. Profit is not assured every year. Prices of the flowers change, yield also changes and there are also climate changes. Economic profitability does not depend on the farm operation itself but on the economic performance of the floriculture sector as a whole.

To ensure economic sustainability, farmers have to incorporate risk management strategies. Some of these include:

- Being able to maintain a low debt to asset ratio.
- Diversifying the flowers planted and grown on the farm.
- Using market-based risk management strategies.

Figure 10.4: Diversifying the flowers planted and grown on the farm is a risk management strategy.

Source: https://krishijagran.com/agripedia/how-to-make-profit-by-protected-cultivation-of-rose/?amp=1

Conservation agriculture can also help improve economic profitability by improving production efficiency. Three major economic benefits can result from practicing conservation agriculture:

- Time-saving therefore resulting in reduced labor requirement.
- Increased efficiency is due to more output for a lower input.
- Reduction in labor costs, machinery operating costs and maintenance.

Budgets can also help improve economic profitability. Predicting prices received, quantities produced and full costs can help create budgets. To make Sustainable agricultural practices profitable among smallholder floriculturists, the government should help provide access to infrastructures, inputs and extension services at affordable prices. Thiscan help facilitate the adoption process of Sustainable agricultural practices among smallholder farmers and help them afford the necessary inputs for a successive adoption of SAPs (Ikerd, 2003).

10.3.3 Most Efficient use of Non-renewable Resources

Generally, it is impossible to use a non-renewable resource sustainably, because one unit used is one less unit from a finite pool. If not managed properly, non-renewable resources can become nonexistent because the rate at which they are used is much faster than that which they are replaced. They include oil, natural gas and coal.

Figure 10.5: Soil is a natural resource that we cannot renew yet we are literally throwing it away.

Source: https://archive.senseaboutscience.org/pages/soil-a-non-renewable-resource-that-were-throwing-away.html

Soil is also considered a non-renewable resource. Any loss of soil and degradation cannot be recovered any time soon. Soil isa very important component of the resources of the land, development of agriculture and ecological sustainability, as a source of food and fiber and a basis for ecosystem services. It is one of the most valuable natural resources (Ikerd, 2003).

Soil, when used efficiently, can reduce poverty, eradicate hunger and food insecurity and it can also stabilize ecosystems. Land degradation needs to be avoided to prevent wastage of this natural resource.

10.3.4 Protection of Public Health

Sustainable flower crop cultivation is linked to public health in many ways:

Climate change – dealing with climate change has become the 21st century's greatest public health problem. It causes heatwaves, water shortages, infectious diseases, floods, respiratory and cardiovascular diseases, malnutrition and even migration. The agriculture sector is responsible for 10% of the EU-based emissions. Climate change can either affect the severity or frequency of diseases that are already affected by climate changes or it can create diseases that are unexpected in places they have not occurred before. It mostly affects those who are poverty-stricken, pregnant women, children, vulnerable occupational groups and immunocompromised (Ikerd, 2003).

Air quality – air pollution causes around 400,000 premature deaths annually. Floriculture produces high levels of ammonia and methane which contribute greatly to air pollution. Air pollution can lead to asthma attacks and other respiratory and cardiovascular health effects.

Agrochemicals – these degrade the ecosystem and pesticides can be an occupational hazard to farmers and any inhabitants of the surrounding area. Pesticides have shown to produce chemicals that disrupt the endocrine system (Ikerd, 2003).

In addition, excessive and improper use of fertilizers can contaminate water sources. High levels of nitrates and nitrites in the water can lead to hemoglobin disorders. Heavy metals such as mercury and lead can also be found in fertilizers and can also seep into the water. When consumed, it can cause kidney, lung and liver disturbances. These can even lead to cancer. Farmers should opt to use organic fertilizers to prevent these issues.

10.3.5 Social and Economic Equity

Floriculture should not only focus on profits but also on relationships that ensure fairness in common environment and life opportunities. By fairness certain categories should be considered: respect, justice, equity and stewardship of the shared world, both among people and in their relations to other living beings (Rikken & ProVerde, 2010).

Work pay forfarm labor is quite low in industrialized countries. Floriculture sectors rely on labor from people who migrate from poorer countries. This leaves farmers vulnerable to changing immigration policies and placing burdens on government social services. Most of these workers accept certain conditions due to their legal status. Some of these conditions are:

- Low pay
- Poor living conditions
- Lack of other job opportunities
- Lack of job security
- Lack of occupational protection

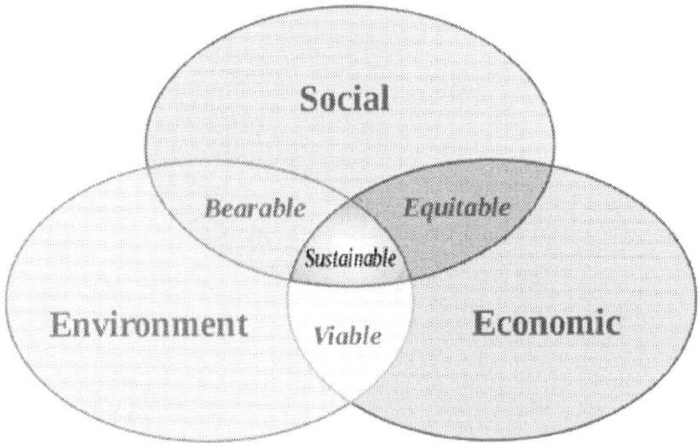

Figure 10.6: This is the principle of Equity and Equality in social and economic justice.

Source: https://en.wikipedia.org/wiki/Equity_%28economics%29

This can be avoided by sharing labor amongst different farms with different flower crops so as to even out the seasonality of work opportunities. Mentoring workers in acquiring skills that allow them to open their own farms, sharing equity in farm profits and even inventing ways to provide affordable education and affordable health insurance for employees and their families. This can help promote economic and social equity (Rikken & ProVerde, 2010).

A sustainable flower farm needs to need o meet the economic and even basic needs of the workers on the farm. Adequate food, water, clothing and shelter should be provided as it is a requirement of social justice. Floriculture should aim at reducing poverty and providing a good quality of life so as to promote economic and social equity.

10.4 INITIATIVES TO MAINTAINING SUSTAINABILITY IN FLOWER CULTIVATION.

10.4.1 Strengthening the Capacity of Flower Crop Cultivation

Capacity development programs help create, strengthen and maintain the capacity of flower crop cultivation. There are certain interventions for strengthening the capacity of flower crop cultivation:

Thinking of flower crop cultivation as a business concept – this helps increase the understanding that a business approach helps improve productivity and livelihood. Training the farmers about the business principles inclusive also of planning and management can go a long way.

Improving functional literacy and basic business skills – illiteracy has a direct effect on the productivity of smallholder farming communities. It is prevalent in rural areas. Having programs that will focus on math, general basic skills and literacy can improve flower cultivation as a business.

Strengthening leadership and governance – the business should run better on democracy. Any new leaders need to be well educated on their roles and on any responsibilities they have. They need to know all the laws required in the floricultural business so as to be the voice for the members.

Figure 10.7: Training the farmers about the business principles inclusive also of planning and management can go a long way in improving flower quality and sustainability.

Source: https://krishijagran.com/agripedia/a-guide-how-to-have-healthy-rose-plantation/?amp=1

10.4.2 Responsible use and Management of Agrochemicals in Flower Crop Cultivation

Improper use of agrochemicals on crops can contaminate water and soils and even lead to poor quality crops. This can affect the health of farmers and any people living in the affected areas. Excessive and improper use of fertilizers can contaminate water sources. High levels of nitrates and nitrites in the water can lead to hemoglobin disorders. Heavy metals such as mercury and lead can also be found in fertilizers and can also seep into the water. When consumed, it can cause kidney, lung and liver disturbances. These can even lead to cancer (Rikken & ProVerde, 2010).

Figure 10.8: Agrochemicals can affect the health of farmers and any people living in the affected areas. Excessive and improper use of fertilizers can contaminate water sources.

Source: https://bioinnovate-africa.org/bio-pesticides-can-secure-our-horticulture/

The first step to avoid these issues is to train the workers in the appropriate way to use agrochemicals. This helps improve their knowledge and competence, especially at the farm level. They should be trained on the acceptable standards and protocols. Responsible use of agrochemicals can improve profitability, the health of the farmers and the workers and food safety. On top of that, it can reduce the impact on the ecosystem.

Agrochemical industries are the main stakeholders. They have the highest responsibility to introduce agrochemicals that are less toxic and hazardous. They should develop and commercialize bio-fertilizers and biocontrol agents that will end up being attractive and alternative strategies. Farmers and the general public should be made aware of the hazardous effect of agrochemicals. Awareness can stimulate pressure build-up for minimum utilization of conventional agrochemicals (Rikken & ProVerde, 2010).

The use of agrochemicals in agriculture has helped improve crop yields and provide protection from pests. However, these chemicals pose a threat to human health, the ecosystem and the environment. It may seem very impossible to completely eradicate conventional fertilizers but we can try to incorporate non-conventional approaches so as to sustain the environment.

10.4.3 Responsible use of Natural Resources and the Reduction of CO2 Emissions in Flower Cultivation

The main natural resource in flower cultivation iswater and soil. Not only are they important in flower cultivation, but also in the functionality and structure of agricultural production systems and for social and environmental sustainability. Agriculture already accounts for 70% of total water consumption globally. If not well managed, it will rise by 20%. Climate change is already causing changes in the water supply. The normal seasonal timing of rainfall has started to fluctuate and there are high occurrences of floods and droughts (Rikken & ProVerde, 2010).

In the case of soil, scientists have proven that fertile land is being lost at 24 billion tons a year due to improper farming practices. Some of these bad farming practices are excessive and improper use of agrochemicals, heavy tilling and multiple sequential harvests.

Figure 10.9: This is a drought-resistant perennial that can be planted ahead of droughts to minimize water use.

Source: https://findingseaturtles.com/drought-resistant-perennials/

To manage this issue a few things can be done:

- Encouraging the protection and restoration of water sources and promote the optimization of water use.
- Promoting waste recycling, and responsible disposal of waste.
- Treating wastewater before reusing or disposal.
- Fostering of soil conservation and improved carbon stocks.

Using workshops and training for farmers can help promote the dissemination of good practices within the floricultural sector.

Agriculture accounts for approximately 4% of the greenhouse gases that are emitted. Manure is one of the significant contributors. There are practices that can sequester carbon and mitigate greenhouse emissions from manure through managing manure to reduce methane and nitrous oxide:

- Covering of manure storage facilities.
- Optimizing manure application to the soil.
- Capturing and combusting methane from manure.

There are also farm practices that can sequestrate carbon by either increasing carbon storage in the soil or prevent the loss of stored carbon:

- Using crop rotations.
- Avoiding over-application of fertilizers.
- Establishing agroforestry systems.

10.4.4 Increased Transparency and Traceability in the Supply Chain in Flower Cultivation

Transparency involves noting and transferring of information along the supply chain. Traceability involves the individual batches of components or purchase orders as they progress through the supply chain. Traceability relies on transparency.

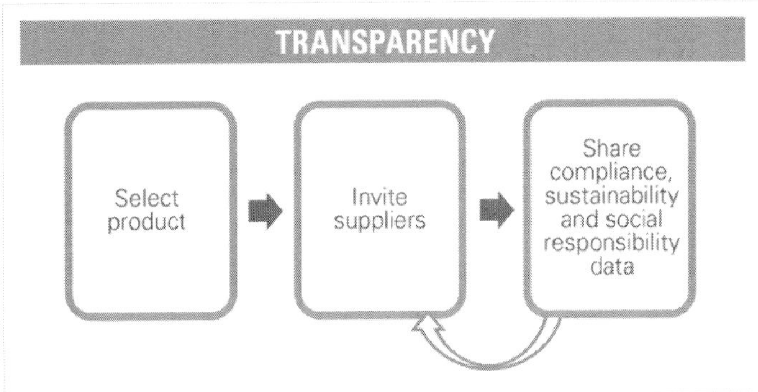

Figure 10.10: This shows the chain of transparency.

Source: https://www.sgs.com/en/news/2018/08/understanding-transparency-and-traceability-in-the-supply-chain

Transparency allows the flower farm to map its entire supply chain, being able to understand and see all the economic operators along that supply chain. The system captures information that is specific to the batch in question relating to each supplier – for example, expiration dates, facility certification, certifying bodies, and even a copy of the certificate (De Pascale and Maggio, 2005).

One cannot implement traceability without transparency. In other words, a farm operation cannot access operational data if it does not fully understand all the economic operators in its supply chain.

Here are the steps to achieving transparency to achieve traceability: (first three steps involve transparency and the last five involve traceability)

- The first step is to select a product.
- Thereafter, suppliers are invited.
- The compliance, sustainability and social responsibility data areshared.
- After transparency is achieved, traceability can be achieved by choosing the specific content to trace.
- Common standards to communicate with the partners are determined such as the batch lot.
- Methods to produce and gather correct data are established and implemented.

- A collaborative platform to store traceability data is then selected.
- Lastly, the determination of how to share data on a platform is established. This requires system development.

10.4.5 Make use of Renewable Energy Sources in Flower Cultivation

Wind, solar and biomass energy can be harvested and these sources can provide a long term solution as a source of energy. This, in turn, leads to a long term source of income. Wind energy on its own can be a source of jobs (it can provide around 80,000 jobs) and 1.2 Billion in dollars as new income for farmers.

Wind can be used to generate electricity and also to pump water. Turbines can be installed on flower farms to provide electricity to the consumers and the electric companies. Biomass energy is produced from organic wastes and plants. Any crops that are grown to provide energy can be grown in large quantities. Corn is as of today they most used source of biomass energy. Switchgrass, poplar and willow are also used and might end up being more popular than corn in the future (De Pascale and Maggio, 2005).

Figure 10.11: Wind can be used to generate electricity and also to pump water. Turbines can be installed on flower farms to provide electricity.

Source: https://www.studentenergy.org/topics/renewable-energy

These crops and biomass waste can be sold to energy companies and also work as fuels in the farm after being converted to energy. This method can provide 20 billion dollars in new income.

Every day, the sun sends large amounts of energy to the earth. Think about it like this, in total, all the energy stored in The reserves likecoal, oil, and natural gas is equal to the energy from the sun for only 20 days. Using solar energy can reduce pollution of the environment, it can save costs from buying conventional sources of energy and it can increase self-reliance. Solar energy can even be used to dry flower seeds and flowers. It can also be used to warm greenhouses (Ikerd, 2003).

10.5.6 Integrated Pest Management in Flower Cultivation

Insects and pests feed on the plant leaves and cause yellowing, destroyed foliage and damaged blooms. These insects and pests can be avoided by pesticides but this can also affect beneficial insects. There are alternative measures that can be used to control insects and pests.

Aphids: These are pests that are shaped like a pear. They suck sap distorting foliage and causing yellowing and falling off of leaves. Feeding on the sap allows them to spread viral diseases to the plant. They also excrete honeydew on the leaves which can make them susceptible to mold. They can be controlled by:

- Strongly spraying the plants with water.
- Natural control by encouraging other insects such as lady beetles and aphid midges which will feed on the aphids.
- Spraying repellents with garlic or hot pepper keep aphids away.
- Insecticidal soap can be used in extreme cases.

Figure 10.12: Aphids sack the sap out of plant leaves and also feed on the leaves leaving them looking very unattractive and it can kill the plant.

Source: https://www.bobvila.com/articles/how-to-get-rid-of-aphids/

Caterpillars: These chew on the flower leaves or along the margins. To eliminate them:

- One can use floating row covers to keep them away.
- One can also encourage other predictors that will eat the caterpillars.

Cutworms: These are wide segmented larvae that can be either black or gray. They are mostly active at night. They feed on the plant's stem and can completely eat through small plants. They can be controlled by:

- Since they are mainly observed in May and June, one can delay planting.
- One can handpick them from below the soil surface.
- Using cutworm collars is important fortransplants.

REFERENCES

1. Adil, J. (2016, May 15). Insect pest of floriculture crops and their control. Retrieved from slideshare.net: https://www.slideshare.net/jawadadil14/insect-pest-of-floriculture-crops-and-their-control
2. Anderson, R. G. (2015). Supplemental Lighting forCut Godetia Production forChristmas and Valentine's Day Crops. Floriculture Research Report, 1-10.
3. Author, G. (2017, April 10). 13 Advantages Of Growing Plants Within A Greenhouse. Retrieved from greenerideal: https://greenerideal.com/guides/13-advantages-of-growing-plants-within-a-greenhouse/
4. Center for Agriculture, Food and the Environment. (2016, August 4). Harvesting and Handling Cut Flowers. Retrieved from umass.edu/greenhouse-floriculture: https://ag.umass.edu/greenhouse-floriculture/fact-sheets/harvesting-handling-cut-flowers
5. Dahal, S. (2013). Post Harvest Handling Of Cut-flower Rose. Department of Horticulture, Institute of Agriculture and Animal Sciences(IAAS), Rampur, Chitwan, Nepal, 18-35.
6. Dait, J. M. (2015). Trends and Issues of the Cut flower Industry ofNueva Vizcaya. International Journal of Novel Research in Marketing Management and Economics, Vol. 2, Issue 3, pp: (25-51).
7. earthcaregreenhouses. (2015, Jan 6). List of Flowers to Grow in a Greenhouse. Retrieved from earthcaregreenhouses: http://www.earthcaregreenhouses.com/blog/news/list-flowers-grow-greenhouse/
8. Eliza Milio, Sky DeMuro, and Orin Martin. (2017). Organic Cut Flower Production on California's Central Coast: A Guide for Beginning Specialty Crop Growers. Center for Agroecology & Sustainable Food Systems (CASFS), University of California, Santa Cruz, 1-11.

9. Everyday Science. (2019, Sep 12). Common Rose Plant Diseases & How To Prevent Them. Retrieved from Agronomy Newsletters: https://cropnuts.com/common-rose-plant-diseases/
10. FAO. (2013). Good Agricultural Practices for greenhouse vegetable crops. FAO Plant Production And Protection Paper, 217-616. Retrieved from johnnyseeds.
11. Greer, L. (2007, August). Growing Cut Flowers Year Round. Retrieved from gpn Greenhouse Product News: https://gpnmag.com/article/growing-cut-flowers-year-round/
12. Henderson, P. (2008). Practical Floriculture: A Guide to the Successful Cultivation of Florists' Plants, for the Amateur and Professional Florist. Massachusetts: Applewood Books, pp.340.
13. Ikerd, J. (2003). Exploring Sustainability in Agriculture. Sustainable Agriculture Research & Education (SARE), 1-6.
14. Jagreeti Gupta and R.K. Dubey. (2018). Factors Affecting Post-Harvest Life of Flower Crops. International Journal of Current Microbiology and Applied Sciences, 7(1): 548-557.
15. Larson, R. A. (2012). Introduction to Floriculture. London: Elsevier, pp.636.
16. Milco Rikken & ProVerde. (2010). The European Market for Fair and Sustainable Flowers and Plants. Trade for Development Centre – BTC (Belgian Development Agency), 11-61.
17. Mohan Bahadur Thapa & Sanjay Dhimal. (2017). Horticulture Development in Nepal: Prospects,Challenges and Strategies. Universal Journal of Agricultural Research, 5(3): 177-189.
18. Mr. Jan Plasmeijer and Mr. Chumi Yanai. (2012). Cut Flowers And Ornamental Plants. Market News Service (MNS), 1-32.
19. newworldencyclopedia. (2016, Sep 15). Flower. Retrieved from new world encyclopedia: https://www.newworldencyclopedia.org/entry/Flower
20. Ravinath, D. (2007). Floriculture: A Viable Business. New Delhi: Excel Books India, 37-87.
21. Reid, M. (2009). Handling Of Cutflowers For Export. Proflora bulletin, 3-25.
22. Singh, A. K. (2006). Flower Crops: Cultivation and Management. New Delhi: New India Publishing, pp.480.

23. Stefania De Pascale and Albino Maggio. (2005). Sustainable Protected Cultivation at a Mediterranean Climate.Perspectives and Challenges. Department of Agricultural Engineering and AgronomyUniversity of Naples Federico II, 29-38.
24. study.com. (2019, Sep 25). Floriculture Jobs: Career Options and Requirements. Retrieved from study.com: https://study.com/articles/Floriculture_Jobs_Career_Options_and_Requirements.html
25. Thippaiah, P. (2005). Floriculture In Karnataka:performance, Problems And Prospects. Agricultural Development and Rural Transformation UnitInstitute for Social and Economic Change, 1-130.
26. USDA. (2019). Floriculture Crops 2018 Summary. National Agricultural Statistics Services, 1-89.
27. Vaughan, R. (2019, Sep 25). CRITICAL ISSUES IN PLANT NUTRITION IN FLORICULTURE. Retrieved from cropnuts: https://cropnuts.com/flower-farming/
28. Vikaspedia. (2019, Oct 11). Gerbera cultivation. Retrieved from http://vikaspedia.in: http://vikaspedia.in/agriculture/crop-production/package-of-practices/flowers/gerbera-cultivation
29. Wallin, C. (2011, March 12). Profitable Plants. Retrieved from profitableplants.com: https://www.profitableplants.com/how-to-start-a-flower-growing-business/

INDEX

A

Abrasion 101
Accumulation 235
Acidified 234
Agricultural worker 78
Agriculture sector 266, 271
Agrochemicals 271, 274
Air quality 271
Ample bed preparation 64
Aromatic quality 148
Asexual propagation 90
Asymmetrical mouthparts 164
Atmospheric temperature 5

B

Bacteria elimination process 130
Bacterial entry 169
Bacterial pathogen 232
Ballerina 29
Bedding crops 124
Bedding flower 28
Biennial flower 143
Biennial plant 63
Biodiversity 267
Black-Eyed Susan 63
Blossom 29, 30, 31, 32, 33, 34, 35, 37, 38, 39, 43, 52, 54
Botanists 112
Bromeliads 46, 47
Bruising 69, 73
Budding 196, 197
Business skill 273
Busseola fusca 167
Busy Lizzie 30, 31

C

California Poppy 63
Calyx 227

Camellias 204
Canterbury Bell 63
Capacity development 273
Capillary moisture 128
Capital investment 142
Carbohydrate content 223
Carcinogenic 48
Cardinal flower 34
Carnations 230, 244, 247
Caterpillars 156
Chemical ecosystem 93
Chemical fluctuations 97
Chemical pesticides 132
chemical spray 6
Childhood labor 79
Chlorophyll synthesis 186, 188
Chloroplasts 236
Christmas star 42
Chrysanthemum gall 158
Chrysanthemums 157, 228, 230
Citric acid 234
Climate change 265, 271
Climate management 114
Climate zone 3
Climbing flower 68
Cold storage 142, 150
Columbine plant 40
Commercial flower farmer 8
Commercial plant 112
Commercial seed packet 93
Condensation 231, 233, 245
Conditioning 233
Coneflower 35, 39
Conservation agriculture 269
Cooling pad 4
Cornu aspersum 161
Corrugated material 122
Cost-effective streamlining 7
Crop cycle 124
Crop disease 171
Crop mites 165
Crop-parasitic nematode 132
Crop propagation 90, 91, 104
Crop residues 268
Crossbreeding 29
Crosspollination 28
Crown gall rot 171
Cultivation system 262
Cultivator 144
Cyclamen 45, 46

D

Deadheading 186, 202, 208, 209
Desiccation 228, 231, 246
Development program 146
Dinance procurement 142
Disbranching 204
Disease management 196
Dissemination 147
Dormancy 91, 93, 96, 98, 99, 103
Dwarf trait 9, 10

E

Earthworms 201
Echinacea 38, 39
Economic performance 268
Economic profitability 269
Economic sustainability 262, 268
Ecosystem service 270
Electricity 142, 150
Embryonic growth 97
Embryos 93, 94, 98
Employment 145, 149
Endocrine system 271
Environmental condition 6, 11, 232
Environmental control 3
Environmental damage 265
Erosion 264, 265, 267, 268

Euphorbia 41
Eye-catching spaces 186

F

Farm manager 77
Fern genera 48
Fertilization 28
Fertilizer 142
Fertilizer mix 125
Fertilizing application rate 92
Filtering radiation 114
Flawless bouquet 73
Floral arrangement 70
Floral businesses 70, 78
Floriculture vegetation 2
Floriculturist 2
floristry 28
Flower crop 28
Flower crop foliage 159
Flower crops cultivation 62
Flower farmer 142, 145, 147, 151
Flower garden 67, 75, 76
Flower germination 66
Flower growing 2, 3, 19
Flowering crop 62
Flowering plant 2, 11, 14, 15, 19, 22
Flowering stem 202, 203
Flower production 6, 7, 17, 23, 142
Flower propagation 90
Flower rhizomes 95
Flower seed germination 91
Fluctuation 76
Fluorescent light 189
Foliage 222, 223, 226, 233, 239
Foliage plant 46, 48
Forced air cooling 230
Foster germination 94
Fragrance 33, 40, 43, 46, 56
Fuller plant 202
Full-sun plants 143
Fungi 224, 226, 231, 233, 234, 238, 239, 245

G

Gardeners 262
Gardening system 120
Geographical 144, 151
Geraniums 32
Germinating plant 156
Germination 188
Gladiolus 228, 230, 250
Grasshoppers 160
Gravitational moisture 128
Gravitational water 193
Greenhouse 112, 113, 114, 115, 116, 120, 121, 122, 123, 124, 125, 126, 135, 137, 139
Greenhouse construction 3
Green solution 45
Ground beds system 117

H

Hardening 233, 243
Hawaiian ginger 75
Helichrysum 157
Herbaceous 2, 11
Hermaphroditism 28
Heteroptera 165
Honeydew 166, 177
Hormone structure 96
Horticultural 142
Household pollutant 51
Houseplant 165
Humidity 189
Hydrotropism 237
Hygroscopic moisture 127, 128
Hymenoptera 159

I

Immature flower 225
Immigration policy 271
Indoor plant 191
Industrial agriculture 81
Integrated Pest Management (IPM) 80
International Labor Organization (ILO) 79
Irrigation 142, 150

L

Land degradation 265, 270
Landscaping project 143
Land sustainability 264, 265
Leaf expansion 188
Leafhoppers 162
Leafminers 166, 167
Leaf variegation 51
Limestone 264

M

Maggots 157
Marketability 142
Market cycle 145
Market data 147
Market demand 262
Market information 147
Marketing cooperative 145
Market product 145
Mass planting 29, 30
Mechanical damage 226, 228, 237, 238, 247, 249, 251
Metabolic disturbance 187
Metathorax 169
Microbe 234, 249
Micro-climate 3
Microclimate control 5
Micro-element 201
Micro-irrigation system 5
Micronutrient 65
Moisture-holding potential 127
Moisture retention 194
Morphological rest 98

N

Natural resource 266, 270, 275
Natural vegetation 160
Nodular sclerite 169
Nutrient concentration 125

O

Occupational-Safety and Health 81
Optimal development 2
Organic content 66
Organic fertilizers 271
Organic matter 64, 67, 193, 194, 199, 201
Organized marketing system 145
Ornamental flora 2
Ornamental herbs 124
Ornamental plant 64, 67, 70, 86, 91, 95, 96, 98, 104, 107
Oxalis species 44

P

Packing density 228
Papaipema cataphracta 168
Parent crop 91, 105, 106
Partial-shade plant 143
Pasteurization 130
Pasturelands 267
Pathogens 156, 175
Pelargonium 2
Percolation 127
Perennial flowers 186, 202
Perennial plant 45, 51
Perfume extraction 28

Pests 156, 157
Petunia 2, 29, 30
Philodendrons 48, 49
Philodendrons originate 48
Photoperiod mechanism 28
Phototropism 236
Physical intervention 100
Physiological latency 98
Physiological resting period 97
Phytochrome 94
Phytoplasma 39
Pigment extraction 28
Pinching flower 201
Plantain lilies 40
Plant bug 165
Plant density 196
Plant development rate 186, 191
Plant environment 222
Plant health 65
Plant illnesses 164
Planting technology 78
Plant kingdom 33
Plant tissue 162
Plastic mulche 114
Pollen grains 28
Pollination 28, 33, 46
Polythene 112, 116, 131
Pot management 123
Pre-harvest procedures 222
Private partnership 146
Productivity 142, 150, 151
Pruning 198
Public health 271
Pupate 167

Q

Quality assurance (QA) 148

R

Raised bed 119, 120
Refrigeration 235
Relative humidity (RH) 230
Renewable energy 8
Renewable sources 265
Reproductive plant 90
Resemblance 32
Revenue 142
Rhizome 47
Risk management 268, 269
Rockwool 194
Rural area 142, 144

S

Sandy soil 66
Sawfly larvae 159
Scabiosa plants 202
Scientific change 147
Scientists 164
Seedbeds 116
Seed coat element 91
Seed development 186
Seedling care 91
Seedling establishment 93
Seedling vigor 93
Seed maturation 93
Seed quality 95
Self-seeding 62, 63
Sexual propagation 90
sexual reproduction 28
Shamrocks 44
Soaking 95, 102
Social media 72
Social service 272
Social welfare organizations 79
Soil basin 92
Soil-borne bacteria 171
Soil borne vermin 131

Soil faster 196
Soilless mixes 194
Soil moisture 127, 130
Soil nutrient 3
Soil preparation 64
Soil quality 265
Soil solarization 130, 131
Soil structure 264
Soil surface 158, 175
Soil texture 66
Standardization 145, 149
Stem expansion 188
Stratification 96, 97, 103
Successive bloomer 143
Sunlight 143
Susceptible floras 75
Sustainability 262
Sustainable agricultural 269
Sustainable production 262

T

Thinning cut 199
Tospoviruses 164
Toxic pesticide 113
Transparent material 112
Transpiration 224, 230, 232, 246
Transportability 143
Transportation 144, 150
Tropical canopy 48
Tropism 236, 237
Turgidity 222, 249

V

Vacuum cooling 230
Variegated cut-worm 156
Vegetable plant 2
Vegetative development 124
Ventilation 76, 190
Vermicomposting 201
Vermiculture 201
Victoria amazonica 112
Vulnerability 195

W

Warm temperature 187
Water intake 222, 227, 234, 238
Water loss 224, 229, 230, 231, 232, 244, 245
Water quality 263, 267
Weeding 196, 218
Wire mesh benches 122

Z

Zygote 28